D0983986

THE WAY AND THE WORD

GEOFFREY LLOYD AND NATHAN SIVIN

THE WAY AND THE WORD

SCIENCE AND MEDICINE IN EARLY CHINA AND GREECE

YALE UNIVERSITY PRESS   NEW HAVEN AND LONDON

Published with assistance from the Louis Stern Memorial Fund.

Designed by Rebecca Gibb. Set in Joanna type by SNP Best-set Typesetter Ltd., Hong Kong. Printed in the United States of America by Vail-Ballou Press.

Library of Congress Cataloging-in-Publication Data
Lloyd, G. E. R. (Geoffrey Ernest Richard), 1933–
The way and the word: science and medicine in early China and Greece / Geoffrey Lloyd and Nathan Sivin.
p. cm.
Includes bibliographical references and index.
ISBN 0-300-09297-0 (alk. paper)
1. Science—China—History.   2. Science—Greece—History.   3. Science, Ancient.
4. Medicine, Chinese.   5. Medicine, Greek and Roman.   I. Sivin, Nathan.
II. Title.
Q127.C5 L6 2002
509.3—dc21
2002003469

A catalogue record for this book is available from the British Library.

The paper in this book meets the guidelines for permanence and durability of the Committee on Production Guidelines for Book Longevity of the Council on Library Resources.

10 9 8 7 6 5 4 3 2 1

To Jacques Gernet and Jean-Pierre Vernant

Contents

Introduction

This book is about the beginnings of science and medicine in early China and Greece. It aims to explore comparison, to find a way of gaining from the joint study of two cultures understandings about each that would be unattainable if they were studied alone. We will explain our working aims and methods in Chapter 1, but a few remarks are in order here about our general goals and some assumptions that underlie them.

As our study proceeded, we found that we were investigating what we have come to call, for want of an established term, a cultural manifold. Rather than comparing concepts or factors one at a time, we begin with the commonplace observation that scientific ideas or medical insights do not occur in a vacuum. They grow in the minds of people with a certain kind of education and a certain kind of livelihood and are inseparable from the rest of their experience. Ideas are part of a continuum that includes what thinkers want out of life, who they consider their colleagues to be, how they agree or disagree with them, how they make sense of the world around them, and what political and social choices they make. Because these are the dimensions of what intellectuals in every culture do, exploring their interconnections is fruitful. Modern scientists try consciously to break or at least hide these links, but in technical enterprises before the mid-nineteenth century, they remain in plain view.

For the sum of all those dimensions we use the term "manifold." Its content is unique to every society, and to some extent to each stratum within it. It is constantly changing, but cultures do persist. We have discovered that we can say a good deal about the Greek manifold that was common to mathematical authors or about the one that Chinese astronomers shared. Delineating a manifold becomes a great deal easier when there is a dissimilar one with which to compare it. We wish to show in this book that that kind of comparative enterprise is not only feasible but illuminating. Rather than comparing concepts or factors one at a time, that is how we proceed.

Anyone who is curious about such matters or who would like to examine Greek and Chinese culture from a viewpoint different from the usual ones is invited to read on. We do not presuppose substantial knowledge of Greece, China, philosophy, science, or medicine. Those who wish to refresh their acquaintance with relevant matters of fact will find two aids: a chronology and an index that identifies concepts and ancient individuals whose work we discuss. For those who wish to examine our evidence or for scholars who would like to carry our reconnaissance further, we provide essential notes and a bibliography.

Some readers unfamiliar with recent studies in the history of science and medicine will not find some of our assumptions obvious. It will be best to state four basic ones. The first three are well established among practitioners of that discipline, though not universally accepted by them.

1. We think of the history of science in the same way that we envision any other species of history. We see it as unfolding from the first tentative explorations, one small step at a time, going in no particular direction, and arriving where we are today by processes that depend on hope, effort (sometimes fruitful, sometimes misguided),

and chance, not on fate or some ineluctable pull exerted by modern knowledge. Each culture began in its own way and made its own path. If no culture, including the Greek, aimed toward modern science, it is idle to ask why anyone, obviously including the Greeks, did not get there. The historical questions that interest us are, rather, In what circumstances did inquiries about the world outside human society begin? and What paths did those inquiries open up? Questions about which of the two cultures discovered more facts or methods similar to today's knowledge tend not only to be distracting but to yield misleading answers. They are misleading because small similarities between past and present are almost always irrelevant to the big picture, and what seem to be striking likenesses tend to fade and disappear under close examination.

2. As a corollary, modern natural science is not the unilinear descendant of Greek natural philosophy. That myth evaporated long ago as historians came to understand the contexts of inquiry. Instead, they trace the ancestry of modern specialties to the cosmopolitan blend of Syriac, Persian, ancient Middle Eastern, Indian, East Asian, and Greco-Roman traditions that formed in the Muslim world. This blend entered Europe beginning about A.D. 1000, bringing many powerful components of which the Greeks had not even dreamt. It stimulated change that has accelerated up to the present day.[1] The simplest way to assemble an adequate comparative account of how science became modern is to ask how people in each of these early technical cultures came to explore the physical world, how individuals proceeded in their own circumstances, what frames of understanding people in each culture created, and how their technical traditions interacted.

3. The research frontier of the history of science, for more than a generation, has moved steadily away from a preoccupation with either social or intellectual history and increasingly toward exploration of the complex realities of which both are parts. There is no convenient label for this comprehensive approach.

Whether it falls under the rubric of cultural history we cannot say, for that term is too vague to specify its own limits. We feel no need for a label.

4. Those who have read extensively in Chinese history (where a bifurcation into social and intellectual history is still common) may be surprised by the absence of Confucianism, Taoism, Legalism, and kindred ideologies from our analysis of thought. Isms of this kind are conspicuously vague. There are at least twenty common senses for "Taoism," and even specialists tend to switch back and forth between them with no apparent consciousness that they are doing so. Isms as used by Sinologues usually refer to beliefs floating in midair, rarely attached to a specifiable group of people.[2] As we point out in Chapter 2, the only long-lasting intellectual lineages in China were the rather diverse ones that transmitted the canon of classics associated with Confucius. For some years, whenever tempted to use an ism, including "Confucianism" (which is problematic in many ways), we have asked ourselves which persons we have in mind, and have discussed them instead. We believe the outcome has been an increment of clarity.

When we met eleven years ago, both of us were already convinced that comparison was too important a historiographic tool to ignore, and had been trying to think of ways to apply it broadly. We both had some rough ideas that we decided to test out together. For several years we fired rough drafts of exploratory essays back and forth across the Atlantic and met in the summers to argue about them. As we took stock of what we had learned, each of us realized that we were looking in new ways at the research fields which we had long cultivated, and were seeing patterns which we had earlier overlooked. At that point we decided to begin work on a collaborative book. Since making that decision, we have reread all of the pertinent primary literature, Greek and Roman on the part of Geoffrey Lloyd and Chinese on the part of Nathan Sivin, with a new set of questions in mind that grew out of our initial explorations. We have also learned a great deal from recent research (much of it generously shared but not yet published) by colleagues who, like us, are reconstructing the wholeness of ancient thought and practice.

The structure of the book is uncomplicated. Introduction and conclusions aside, we devote two chapters to China, on the one hand, and two to the Hellenic, the Hellenistic, and, to some extent, the Roman worlds, on the other. We investigate, for reasons that we explain in Chapter 1, the six hundred years from 400 B.C. to A.D. 200. We do not try to devote equal attention to each era within this period; our coverage is limited to what our larger argument requires. We say "requires," for our desire to sketch a complex train of events briefly for nonspecialists leaves room only for what is essential. It also rules out arguing here with colleagues whose conclusions on contentious issues differ from ours.

The chapters on Greece, like the chapters on China, tell a single story, which begins with livelihood, at the social end of the cultural manifold, and moves along to scientific and medical concepts, at the intellectual end. The accounts are each divided between two chapters to make the comparisons easier to follow.

We have followed several conventions. First, in writing about thinkers of European antiquity (as they presented themselves to their colleagues and rivals), we have found Marcel Detienne's celebrated expression "maîtres de vérité" so serviceable that we have adapted it for more general use as "Masters of Truth." As a complement, we refer to Chinese masters from time to time as "Possessors of the Way." The implied comparison is too complex to reduce to one of yin and yang.[3]

Second, we use the Wade-Giles romanization for Chinese rather than the Pinyin system. This choice was not made on the basis of linguistic merit; neither system has much to recommend it. Using Wade-Giles means that readers who have not studied Chinese will find it easier to compare our account with its most important predecessors.[4] It also means, unfortunately, that our study will be less compatible with publications on modern China, for which Pinyin is now universal. Because such decisions also have contemporary resonances, it probably also means that some arbiters of such matters will accuse us of political incorrectness. A desire to align ourselves with the political right or left played no part in our choice. Fortunately, the discipline of Greek and Roman history does not impose such loyalty tests on its practitioners. The index provides Pinyin equivalents of personal names.

Third, the translation of official titles remains quietly controversial in Sinology, though a matter of negligible concern in Classics. At least three systems are in current use for Han China (between 200 B.C. and A.D. 200). We follow with only a few exceptions that of C. O. Hucker, because he set out to find translations that cover as much as possible of Chinese history.[5] We also translate the titles of emperors and the names of reign eras because they were chosen precisely for their meanings. Clumps of transliteration would make those meanings inaccessible.

Finally, all translations, unless otherwise stipulated, are our own.

*   *   *

Although we can in full confidence claim responsibility for our errors of fact and understanding, we have benefited from the suggestions and criticisms of many colleagues, far too many to express our gratitude to every one.

We are indebted to all those who attended several conferences in which our work in progress was discussed. These include the Seventh International Conference on the History of East Asian Science, Keihanna, Japan, in 1993; two interdepartmental faculty seminars organized by Jeffrey Rusten at Cornell University in 1993 and 1995, which made it possible to discuss our work with experts in many fields when it was still very tentative; a session in the conference "The Understanding of Nature in China and Europe from the Sixth Century B.C. to the Seventeenth Century" organized by Gunther Dux and held at Rheine, Westphalia, Germany, in March 2000, devoted to an early draft; a conference organized by Benjamin Elman at the Institute for Advanced Study, Princeton, in April 2000 to discuss parts of the manuscript; and a three-day international meeting of scholars at the Needham Research Institute, Cambridge, England, in July 2000 to critique the whole draft. We also acknowledge the contributions of those who attended the many colloquiums and seminars on the project that we presented together and individually in Asia, Australia, Europe, and North America. We are grateful for written comments on the manuscript from Karine Chemla, Jacques Gernet, Michael Loewe, Shigeru Nakayama, Michael Nylan, and Robert Wardy, and from Nathan Sivin's students Philip Cho, Aisha Lyons, and Carolyn Naylor. We also wish to acknowledge the detailed criticisms by Vivian Nutton and Heinrich von Staden, as well as help from Asaf Goldschmidt, Sun Xiaochun, and Mary Beyer. We wish to thank Mary Pasti for editorial services well beyond the call of duty.

We wish finally to express thanks for grants to Nathan Sivin from the Research Foundation of the University of Pennsylvania (1992–1993), its Center for Chinese Studies (2000), its School of Arts and Sciences (2000–2001), the European Association for Chinese Studies (1995), the Chiang Ching-kuo Foundation for International Scholarly Exchange (1998–1999), and the American Philosophical Society (2000). The conference at the Needham Research Institute was generously supported by the British Academy, the French Embassy at London, and the Henry Arthur Thomas Fund.

THE WAY AND THE WORD

1 Aims and Methods

China and Greece were two of the ancient civilizations where people began to raise fundamental questions about a wide range of phenomena and to answer them on the basis of systematic investigations. Among the subjects that interested them were the movements of the heavenly bodies, the workings of the human body in health and in illness, the different kinds of animals and their behavior, the varieties of plants and their properties, the relations between harmonious sounds and what distinguishes them from discordant ones, and the nature of the changes that physical objects undergo. People today tend to think of those investigations as belonging to science and as contributing to its early development. But such judgments can be misleading. The ancient investigators had no idea of what science was to become, and did not even have a category that corresponds to modern science. What, then, did they think they were doing and why?

The fascination of comparing the civilizations where these investigations began lies in the mixture of similarities and differences they present. Four general similarities stand out. The first and most fundamental is that Greece and China evolved comparably elaborate cultures, with languages and abstract conceptual structures that could be used to explore every aspect of individual and collective experience. As is always the case, language incorporated constraints on expression, but it kept evolving to let people express whatever they freshly observed or reasoned out.

The second basic similarity is that people in both societies saw the need for such inquiries in the first place. They were not content simply to accept, as the last word, the set of beliefs that tradition or convention handed down.

Third, in both societies specialist groups took the lead in many branches of study, gaining authority to acquire, present, and interpret new knowledge or understanding.

Fourth, people in both ancient societies undertook the studies they did in the conviction that what they learned was needed to understand where humans fit in the universal scheme of things and that this understanding would help them to order human affairs. In both societies, in other words, such investigations were intensely value-laden.

Within each of these broad similarities, there are important differences. Although both societies used, and went beyond, tradition, the ways they responded to the legacy of the past differed. So, too, did the roles and the ambitions of those who did the investigating, as well as the conditions under which they worked. Although the studies in both cases were steeped in values, the values they exemplified were far from identical.

In addition to the similarities in the subjects that Chinese and Greek investigators studied and in some of the methods they used to study them, they were faced with analogous difficulties in a further respect. They had to persuade their contemporaries, or some group of them, that their inquiries were worthwhile and their results valid, especially when their ideas broke with or criticized traditional beliefs. The techniques of persuasion they used, the modes of presentation of their results, the audiences whose opinions counted most, all interest us. The differences here, too, underline the point that ancient inquiries did not all follow the same pattern. This is a further respect in which it would be quite misleading to think in terms of a single linear progress toward modern science. Chinese and Greeks shared a desire to increase understanding but had different ideas about how to go about it.

Many scholars have written on the investigations into physical phenomena made in one or the other of our two ancient societies, and some have made comparisons between them. We have learned from our predecessors' studies, but our own approach differs fundamentally from theirs in one crucial respect. The key notion that guides our work is that the intellectual and social dimensions of every problem are parts of one whole. This is not a new idea. But we make it central to our investigation.

What we are studying in each culture is what we term a manifold. It comprises the inquiries that interest us, not only their intellectual, social, and institutional dimensions but also the interaction that unites all of these aspects into a single whole. We regard as aspects of the whole how people make a living, what their relation is to structures of authority, what bonds connect those who do the same work, how they communicate what they understand, and what concepts and assumptions they use. We do not think of social factors as determining thought nor of ideas as changing society. These are not external causes. Thinkers respond to, but also influence, institutions and prevalent values. Thus we do not speak of inquiry in context. Context is not an autonomous setting that may or may not be connected to inquiry. Technical work and its circumstances are parts of one thing, even though the specialization of modern scholarship encourages dismembering it. Our effort to encompass that unity of Chinese and of Greek thought about the physical world has led us toward a new and, we hope, productive approach to comparison.

We have confronted other equally fundamental questions. What do we mean when we speak of ancient science? On what basis do we decide that phenomena in ancient cultures are comparable?

What Is Ancient Science?

Defining today's science is in one way straightforward. We merely need to specify academic degrees, employment in research, publication in technical journals, professional licensing, and other criteria that identify specialized communities. What criteria can one apply to a time before any of these existed?

The answer has two parts, first a concession and then an elucidation. Obviously the transformations of science in the past two centuries completely overshadow all earlier changes. Judged from that standpoint, it would be wise to admit that in the fullest modern sense there was no science at all before 1800 or so. This is a matter of lacking not only the explicit concept but also the institutional frameworks of modern science—the research laboratories and university faculties devoted to its pursuit.

Yet in the ancient world, people already entertained ideas about the stars, the human body, the variety of living beings, the composition of things and the changes that they undergo. Inquirers attempted, in other words, to understand the world that lay outside the social realm, and some thought hard about how to do so. We will use "science" here as a conventional placeholder to cover such studies as these. The mark of science, in that usage, lies in the aims of the investigation and its subject matter—the bid to comprehend aspects of the physical world—not in the degree to which either the methods or the results tally with those of later inquiries, let alone modern science. As for "scientist" and similar words such as "astronomer" and "cosmologist," we use them simply as shorthand designations for those who engaged in the activities explored here.

To treat premodern science as a mere composite of these studies does not resolve any of the problems. It merely shifts the focus to what each of them comprised, to what passed as astronomy, or medicine, or zoology, and so on. But the approach does have one immediately salutary effect, namely, putting us on our guard against generalizing across all those domains. What may be true of any one of those fields may or may not apply to others, even within a single period and in the same society. Indeed, each ancient culture developed answers to the important questions in its own way.

A word about the titles by which the ancients identified practitioners will illustrate the need for caution. Some labels pick out specialist occupations, such as doctor or healer, although "healing" included an enormous variety of practices in both China and Greece. Two other terms to be wary of are "astronomer" and "mathematician," both derived from Greek roots. Greeks who went by either title were as likely to be engaged in casting horoscopes as to be students of astronomy or mathematics in any modern sense. There is, paradoxically, an advantage in the relative unfamiliarity of the two main Chinese terms for the study of the heavens, li-fa, which covers methods of making ephemerides and other computational tasks, and t'ien-wen, the investigation of the "patterns in the heavens," including cosmography, observation, and the interpretation of omens. The larger lesson to be learned is that how the ancients themselves defined their subjects is the best place to begin—though not necessarily to end.

Two of the most general terms, one Chinese, the other Greek, serve as reminders that the ultimate goal of investigation, when construed as wisdom, was equally problematic in both societies. The Chinese used the term hsueh, "study," for this pursuit. For most early thinkers, "study" was as much a moral as an intellectual enterprise. Its aim was not just to learn facts or develop cognitive skills but to shape one's life. The goal of self-cultivation—spiritual, mental, physical—was sagehood.

"Love of wisdom" is the basic sense of the Greek term philosophia, from which the English "philosophy" is derived. But what that truly consisted in was as controversial in ancient Greece as it has been ever since. Self-cultivation, there too, was one possible component. At the other end of the spectrum, some Greeks who saw themselves as philosophers were principally involved in teaching the skills of public speaking. But among the varied subjects that self-styled philosophers often took up was the study of the physical world and the cosmos as a whole.

That is the sense that is important for the kind of "philosophy" with which we are concerned here. We will use the term "philosopher" conventionally in talking about Greek or Chinese thinkers who investigated not only society but the world outside it. Other topics that have become branches of modern philosophy, such as logic and moral philosophy, are not our central concern. We will study how thinkers in both cultures construed the relations between the macrocosm and the two microcosms, society and the human body (Chapters 4 and 5). We will see that the basic concepts the Greeks and the Chinese used to articulate their ideas differed profoundly, the Greeks focusing on nature and the elements, the Chinese on ch'i, yin-yang, the five phases, and the Way. This raises our next fundamental question.

What Is Comparable?

The most fruitful comparisons begin not with individual concepts or methods but with complexes of thought and activity seen in their original circumstances. Whether it is a matter of studying the stars, or the human body, or harmonies, or the cosmos as a whole, the first step is to analyze what the ancient investigators themselves say they were trying to do—their conception of their subject matter, their aims and goals. That evidence is variable, to be sure, for on these topics some writers are far more forthcoming than others.

When we have no explicit evidence on the writers' own perceptions of their inquiries, attempting to infer what they thought from what they did is fraught with difficulties. That is clear from, among other things, the mismatches we can identify between aims and practice when we have explicit evidence on the former. Aristotle is a case in point. His conduct of zoological investigations hardly tallies with all his own programmatic statements about how to proceed.

Understanding the ancient writers' views about their subject matter takes us into an altogether more promising field than counting up their accomplishments by modern criteria. It leads to such questions as the following.

What made up the study of the heavens, the earth, or their parts, and what made these inquiries important?

What did the living human body contain? How did it work normally, what disrupted its workings, and how could a layman or a healer overcome the disruption? How did people go about answering these questions?

What kinds of description, prediction, explanation, and demonstration did investigators pursue? How did they use them?

Who were the investigators? What motivated them? Did they agree on aims? How were they recruited and trained? To what extent were their educations formal? What were their relationships with their teachers? How did they align themselves with political and social authority?

Did they use tradition to legitimate their ideas? When and why did they reject its authority?

Did scientific writing accept and elaborate popular beliefs, or did it oppose them? How did it justify its departures from them?

Within what institutional structures (if any) did the investigators operate? Did they belong to recognized occupations? How did they make a living, or didn't they need to do so? Why did employers, patrons, or governments support them, and what did they do in return for such support?

The earlier the period we study, the more likely it is that inadequate evidence and bias in the sources will thwart us. A comparativist approach may compound the difficulties. The sources are of more than one kind and in more than one language. A still greater challenge lies in the need to move between, and compare, whole conceptual schemata. Yet if the comparative history of ancient science is peculiarly demanding, it can also be especially rewarding.

The chief prize is a way out of parochialism. What we stand to gain is a better understanding of early science as a phenomenon that took different forms, equally interesting, in different societies. No one society monopolized the desire to know about the movements of the stars or the workings of the human body, or to extend that knowledge by deliberate inquiry. Scholars whose work is confined within a single cultural area easily suppose that its ways are natural and inevitable. Looking across the borders at other cultures or traditions reveals how mistaken that may be.

People familiar only with traditions of European science naturally assume that physical thought could not have evolved through several early stages without some notion of elements. Greek element theories claim that things are composed of basic constituents that do not necessarily resemble what they constitute. This claim built on the idea that reality is hidden at some deeper level than human senses can apprehend. But that fundamental claim had no counterpart in China. Chinese discussed change in terms not of rearranging basic materials but of the dynamic mutation of a unitary ch'i, which they sometimes analyzed as two complementary, opposed aspects of a process in time or configuration in space (yin and yang), or sometimes as five aspects (wu-hsing, "five phases"). Wu-hsing used to be mistranslated as "five elements," but it corresponds to neither classical nor modern concepts of elements. The error goes back to the Jesuit missionaries in seventeenth-century China, who were intent on showing that the indigenous doctrine was inferior to the four-element theory that they were teaching.

Just as starting with European preoccupations is bound to distort the understanding of Chinese science, the converse is equally true. There is no justification for assuming a counterpart in Greek physics or medicine to the Chinese notion of *ch'i*, which is not only the material stuff in everything but the vital energy that makes it possible for things to grow and change, and the fine, essential matter that is the vehicle of consciousness.

Methods

We have already discussed the chief ingredients of our method, but it may be useful to summarize them. We are not comparing things or concepts but whole processes. We look at ideas, their uses, the social interactions that elaborated them, and their adaptation to state power as dimensions of a single phenomenon. We try to reconstruct how people at the time understood their own practices and concepts, rather than how authors of modern textbooks would evaluate their work.

We do not assume that ancient science developed, or should have developed, in just one way. We do not allow either the Chinese or the Greek experience to determine our agenda. Rather, we explore complexes of similarities and differences in both to throw light on how each society articulates its experience. Only by comparative studies, we submit, can such correlations be reliably established. The ambitious aim we have set ourselves is to explain why the various sciences that the Chinese and the Greeks developed took the form they did.

Period

We concentrate on the rich and challenging materials for China and Greece from around 400 B.C. down to approximately A.D. 200, that is, before Buddhism became a powerful influence in China and before Christianity became dominant in the Greco-Roman world.[1] In the same period, of course, two cultures may have little in common. It so happens that both China and Greece passed through analogous transitions and left records of comparable richness in these six hundred years. There is no particular reason why this should have been so. By fortunate accident it is.

We can confront Euclid and Ptolemy with several Chinese classics: *Mathematical Methods in Nine Chapters*, the *Chou Gnomon* (a treatise on mathematical cosmology), and the two earliest Chinese astronomical treatises. In medicine we can set Galen and his predecessors side by side with a series of sources, from the Ma-wang-tui manuscripts, buried in a tomb that was sealed in 168 B.C., to the *Canon of Eighty-One Problems*, probably written in the second century A.D. There is, however, no Chinese corpus of work equivalent to Aristotle's writings on physics or animals or those of Theophrastus on plants. Important Chinese writings on other topics have no Greco-Roman counterparts—for instance, those on resonance as an explanation for physical interaction.[2]

Many documents also survive from these six centuries that bear on the questions we have raised about Greek and Chinese cosmological thinking. We can explore the relationship between Aristotle and his predecessors and the ongoing dialectical debates between the main rival philosophical schools of the Hellenistic world: Platonists, Aristotelians, Stoics, Epicureans, Skeptics. Again in China there were analogous transitions, when early ideas were redefined and synthesized. These we typify in three important works that date between the mid-third and mid-second centuries B.C., the *Springs and Autumns of Master Lü*, the *Book of the King of Huai-nan*, and *Abundant Dew on the Spring and Autumn Annals*, and then in two further books composed around the turn of the millennium, Yang Hsiung's *Supreme Mystery*, and the *Inner Canon of the Yellow Emperor*. The last work, though devoted to medicine, is the most important source for Chinese cosmology in the last half of our period.

Historical Setting

Historical periodizations are always more or less artificial. The conventional boundaries in Greco-Roman history between the classical, Hellenistic, and Roman periods mark cultural shifts more pervasive than do the dynasties normally used to periodize events in China. We will attend to the important political changes throughout the course of our analyses in Chapters 2 to 5, but some brief remarks here will set the scene for later discussions.

In the classical period in Greece there were many autonomous city-states, whose political arrangements, over time, varied from full-scale participatory democracy through oligarchy and constitutional monarchy to tyranny. The two most powerful were Athens and Sparta. Estimates of the sizes of populations involve guesswork, but specialists have estimated that in the fifth century B.C. Athens had a population of less than a quarter of a million, perhaps half of them citizens and their families and the other half slaves and resident free aliens (*metics*). Most city-states were appreciably smaller; even in Sparta the full citizens numbered a mere eight thousand male adults in the 480s B.C. Firm numbers are hard to arrive at, but the order of magnitude suggests an important contrast with China, which in the census of A.D. 1 to 2 had more than a thousand county-level units and a population of nearly sixty million. These estimates are perforce nearly five hundred years apart. It is impossible to offer reliable figures for China earlier or for the Hellenistic world.

The end of the classical period in Greece is marked conventionally by the conquests of Alexander in the 330s B.C. On his death, in 323, the territories he had conquered—from mainland Greece in the west to modern-day Afghanistan and Uzbekistan in the east—were divided among his generals—Seleucus, based in Babylonia; Ptolemy I, in Egypt, and so on. This meant an end to the effective political autonomy of the classical city-states, although some, Athens especially, retained considerable cultural prestige. However, as we will see in Chapter 3, the first three Ptolemies engaged in an ambitious program to make their capital city, Alexandria, rival Athens as a center of learning and art.

Rome's conquest of Egypt in the year 30 B.C. marks the third of our three periods. In this one the seat of power shifted from the Hellenistic kingdoms to Rome, but with some continuities in the cultural domain. Athens remained the chief center of philosophical activity, and Alexandria continued to attract important mathematicians, astronomers, and medical theorists. Although many members of the Roman elite were suspicious of Greek learning, some (such as Cicero) went to Athens to be taught philosophy and rhetoric. Movement in the reverse direction, of Greeks going to Rome to further their careers, grew in importance only at the end of our period. Galen is an example. He was educated in both philosophy and medicine in cities in the Greek east (including a visit to Athens and study at Alexandria), but he came to practice in Rome and eventually became court physician to the emperor Marcus Aurelius.

In China, too, important shifts took place in society and politics as well as in thought. The conventional periodization according to dynasties obscures rather than clarifies some of these. We begin, in any event, at the fag end of the Chou dynasty. In the Warring States period (480–221 B.C.), seven large states proceeded to gobble up seven lesser ones and then each other. By 256 the state of Ch'in was able unilaterally to abolish the Chou dynasty, although its victory over its rivals was not complete until 221. Movement toward unification was well under way by 400 and continued in the Han dynasty down to 156. The collapse of the Ch'in dynasty in 206 B.C. did not at the outset mean peace or a single locus of power. The new Han rulers had to achieve both. In this era of gradually decreasing turmoil and much institutional innovation, social mobility was high. The emperors largely drew on the Ch'in's ritual and theoretical justifications for monarchy rather than creating new ones.

Then came a period of consolidation and expansion. The reigns of two emperors from 156 to 87 B.C. were prosperous, full of grand projects that carried the sovereignty of the Han to Central Asia and Vietnam. The government replaced pre-Han institutions with highly integrated ones that made full use of the elite. It supported an ideology that drew on many different currents of Warring States thought to argue that the order of the good state (that is, the Han) mirrored that of the macrocosm.

Over the century of decentralization, from 87 B.C. to the restoration of the Han in A.D. 25, more and more of the land slipped into the hands of local magnates who could avoid paying taxes. Increasing tax rates for ordinary farmers eventually drove them into the hands of the great landowners. Wang Mang, at the turn of the millennium, is represented in traditional dynastic history as a usurper. He is better seen as a reformer unable to curb the accelerating growth of independent estates as the imperium fell apart.

The last phase, collapse, followed a promising restoration. The new Eastern Han was well poised to create a new deal. But the emperor and his most powerful generals, themselves owners of great landed properties, could not resolve the perennial tension within every member of the elite between his obligations to the state and those to his clan. By A.D. 57 half the population were off the tax rolls, and the remainder were suffering keenly. Within the palace the intrigues of corrupt officials, the families of imperial wives, and eunuchs (despised personal tools of the emperor) left less and less scope for effective government. In the 160s the eunuchs, who had a decisive grip on appointments and every other facet of power, imprisoned and killed thousands of officials and their adherents, peaking in the mass slaughter of 168. In 184 massive local rebellions precipitated a final succession of catastrophes.

From the standpoint of writing on science and medicine, the first of these four political stages, unification, motivated the *Springs and Autumns of Master Lü*, which contains important technical discussions of science and medicine. Practically oriented mathematical, medical, and divination texts based on natural phenomena and calendrical cycles from this period have been excavated in recent years. Other important works of synthesis and practical technical writings appeared in the second stage, consolidation and expansion. In the third, decentralization, the first substantial scientific and medical classics appeared, and in the fourth, collapse, they were widely enough disseminated for their elaboration and annotation to get under way.

The formative interactions of ideas and political circumstances will become clear in the detailed examination of Chinese institutions and science in Chapters 2 and 5. The interplay between political changes and cosmological writings in the Greco-Roman world discussed in Chapters 3 and 4 is not as tight. Still, at the most general level there, too, there was a shift toward domination by larger power blocs and eventually by the Roman empire, and certain transitions in philosophical ideas, notably an increasing sense of the authority of the past.

The political scene and the involvement in it of philosophers and scientists show two salient features. The first relates to scale. Classical Greece was a patchwork of city-states, each with territories and populations that were minuscule by the standards that generally apply to China. The difference bears on the practicality of political experimentation in some of the modes we find in Greece. The effective working of the full democracy, for instance, depended on the entire citizen body's access to the Assembly. The size of China ruled out any such experiment, even had someone conceived it.

Governments and even constitutions in classical Greece were much more diverse and changed far more rapidly than was the case in China. Some Greek philosophers may have been just as eager as their Chinese counterparts to advise those in power. But in the democracies especially, power lay in the hands of an ever-shifting majority. In the move from political diversity to unification, the involvement of Greek intellectuals in the entourages of Hellenistic, let alone Roman, rulers was appreciably less than that of their Chinese contemporaries in imperial politics. Some of the latter intellectuals produced ideas that legitimated and justified unified monarchic rule in China.

We will return to the issues that these observations suggest at the end of our study. In the next two chapters we will review the social and institutional frameworks of Chinese and Greek philosophy and science, and in the pair that follow, the fundamental concepts that Greek and Chinese thinkers produced.

2 The Social and Institutional Framework of the Chinese Sciences

Social Distinctions

What did it take to become a philosopher, scientist, or physician in ancient China and Greece? Did it depend on what stratum of society one came from? How did those who took up these endeavors make a living? Did that affect the inquiries they pursued and the way they pursued them? What part did their technical work play in their careers? Do the answers to all those questions vary according to the period or the discipline? And how do they differ in Greece and China?

The most fundamental difference in early Chinese society was between those who were eligible for office and those who were not. As Mencius put it, "Some labor with their hearts and minds; some labor with their strength. Those who labor with their hearts and minds govern others. Those who labor with their strength are governed by others." In 400 B.C. this gap was a matter of birth. Even in A.D. 200, the sons of farmers did not ordinarily become high officials.[1]

Even before the six centuries we are studying, a common label for those whose livelihood depended on their hearts and minds was "*shih.*" The evolving meanings of this label as society changed are very much to the point. In the eighth century B.C. it referred to the lower strata of hereditary aristocrats entitled to bear arms. They expected each other to be more or less literate but not learned. In the endless political convulsions from that time on, peasant armies on foot took over the fighting from wellborn warriors in their chariots.[2] The nobles who inherited ministerial posts dropped out of the local courts as civil servants subservient to the rulers took their place. Because members of bureaucratic families regularly inherited office, their clans formed a new, mostly lower aristocracy. As wars wiped out state after state, and ruling families and powerful rivals struggled within states, the losers lost their status. "*Shih*" came to designate all sorts of wellborn men, no longer bred to fight, no longer heirs to power, supporting themselves by official employment, patronage, and other pursuits that required literacy or other expertise. As some fell in the world, others rose from obscurity. The result was greater social diversity in a no longer entirely closed (but far from very open) elite.[3] This change is apparent in the most important philosophers of the late Warring States era, who varied from Mo-tzu, putatively a condemned criminal, to the prince who wrote the *Book of Han-fei-tzu*.

Confucius acknowledged an early stage of this transformation. He redefined "*shih*" to imply literacy, learning, self-cultivation, a life centered on ritual, a mind responsive to it, principled action, and a yearning for public office. He also seems to have admitted descendants of artisans and merchants to the ranks of his disciples, although his successors did not generally follow this pattern.[4]

The Han dynasty changed the equation by instituting a central civil service, but authors continued to use "*shih*" for the pool of those potentially qualified to join it. By 100 B.C. (four centuries after Confucius), *shih* were likely to be landowners, wellborn but seldom titled and usually literate. Those whose forebears had fought for the Han at the outset of the dynasty often inherited local posts, but these carried far less prestige than a palace appointment. By A.D. 200 shih tended to come from wealthy families (now wellborn by definition) and to be educated in the classics; many disengaged themselves from the failing central government.

Given all these shifts, a modern reader often cannot be sure what an author meant by "shih." "Gentleman" is a suitably broad translation. The English word, too, originally designated "one who is entitled to bear arms, though not ranking among the nobility." It thus usually referred to the son of a titled family who did not inherit, and to his descendants as long as they remained genteel. Like "shih," it weathered the transition in meaning, first to "a man in whom gentle birth is accompanied by appropriate qualities and behaviour" and then to, "in general, a man of chivalrous instincts and fine feelings." The shift of emphasis from behavior to thought and emotion fits the Chinese case nicely.[5] In this book, when a source emphasizes the services that shih gave as officials or clients rather than their social standing, we render it "gentleman-retainer." "Elite" indicates people qualified by birth and education to hold office, whether or not they did so and whether or not the changing meaning of "shih" applied to them.

Chinese authors from the fourth century B.C. on often used a schema of gentleman (broadly defined to include holders of power), farmer, artisan, and merchant (in top-down order) to suggest a firmly established social hierarchy. Despite the great complexity of Chinese society, elite norms—unlike realities—discouraged occupational specialization. If farmers were self-sufficient, there should be no need for artisans and merchants. Nevertheless, the four-level schema never encompassed the variety of divisions in society; more than a dozen levels occur in early speeches. But the four became standard when discussing how to maintain hierarchy.

Typical is a story in Narratives from the States (ca. 306 B.C.) in which the powerful Duke Huan of Ch'i and his sage advisor Kuan Chung discuss the "ways of the former kings." One of these ways was to prevent the "four kinds of people"—in the order noted above—from living together, lest social mixing make them disorderly in speech and behavior. The ancient sage-kings, Kuan claims, made *shih* live in "pure" surroundings, artisans in official buildings, merchants in market centers, and farmers in the countryside. According to him, each group spontaneously maintained its own social order. Kuan proposes reviving this gradation. Other sources, each for its own purpose, discuss the "four kinds of people of ancient times," not always in the same sequence.[6]

This gradation was plainly meant to describe an archaic society so small in scale that each group could police itself. Such discussions are typical of codes that portray a perfect order in the past. What they have in common is nostalgia for an imagined hierarchy—a segregation—that has already slipped away but which, they argue, a ruler can revive by fiat in order to stop undesirable change.

This schema was, in other words, part of an unending governmental quest to control society from the top down—a kind of control that was out of the question in the Greek city-states. Plato proposed something analogous in the *Republic* and the *Laws*, but even his own successors did not build on either of his blueprints. Chinese bureaucracies again and again set forth Kuan's totalitarian dream (which Confucius, though no democrat, did not share) in concrete legislation and policy. The functionaries never succeeded, but they never gave up. Their policies accomplished with fair consistency the little they could before the twentieth century—namely, to educate members of the small elite to behave conventionally, say the right thing in public, and feel ashamed or isolated when they thought the wrong thoughts. That much is what the classics made possible.

So much for ideals. What were the actual social fissures?

Wealth was regularly changing hands. With a landowning, office-holding elite eager to consume and purveyors ready to oblige, its redistribution was inevitable. In the second half of our period, members of families who had made large fortunes from trade or manufacture were becoming officials, and scions of clans with traditions of civil service were going into business. Independent small farmers were unwillingly becoming tenants on large estates exempt from ruinous taxation. The traditional boundaries between the social strata were thoroughly blurred.[7]

The conventional key to status and thus privilege remained education, and access to it continued to depend largely on birth. Most candidates for official posts were recommended by local officials; others, as relatives of high civil servants, were entitled to appointments. The new elite that evolved after Confucius's time took literacy as the mark of the gentleman. At first it took not much more than basic reading and writing to reach the mark. When the state began from the second century B.C. on to make classical learning a qualification for some civil servants, the standard eventually rose to require memorization of large ancient texts. Mastery of rites and social forms, which depended largely on upbringing, was just as important a qualification. When recommenders submitted formal evaluations of virtue, as they were expected to do, it was the exemplary comportment of the well-bred, rather than the internalized moral values of the orthodox classics, that they usually meant. Until the seventh century A.D., examinations were not a vehicle for mobility between classes, and even afterward they carried few of the lower orders to glory.

A bookish family at a given time might not have a member in office or might lose its land. Poor *shih* families understood that education kept them respectable. Literacy became a more desirable resource as the guaranteed advantages of aristocratic birth bulked smaller. Learning remained out of reach for most commoners, however intelligent or ambitious.

Within occupational groups, the Greeks sometimes asserted literacy as a mark of prestige (see p. 87). But in China by 400 B.C. it was too common among the elite and too sparse outside it to signify exceptional status. What played the corresponding role was fastidious ritual behavior and moral sentiments—marks of membership in the governing class that others, educated or not, could emulate.

This comes out in a famous anecdote of the early Han in which two high officials encountered Ssu-ma Chi-chu, a genteel diviner plying his trade in the open-air Eastern Market of the capital, and introduced themselves. "Seeing that they appeared to be learned men, he treated them with polite formality, having his disciples invite them to be seated." He continued to instruct his pupils, drawing fine distinctions in cosmogony, cosmology, ethics, and prognostication. "In the thousands of words he spoke, every one was concordant with good sense." His visitors, impressed, asked him how someone so cultivated could be "living so degraded a life, practicing an occupation so unwholesome." Laughing, he took the high ground. "You great gentlemen seem to be of the sort who have attained the arts of the Way. How, then, can your question be so vulgar and crude?" He declared that honest, frank people were seldom found alongside the sycophants in the upper ranks of officialdom. The mark of a sage, he insisted, was not position but integrity. What made Ssu-ma sagely was not his learning, impressive though it was, but his conduct, grounded in the rituals and moral principles of high antiquity. As a *shih*, or nearly so, the diviner was showing that elite values were not confined to officials. Society offered mobility to only a few of the fortunately placed and exceptionally talented; still, there was considerable fluidity along the borders between classes. Ssu-ma never became an official, but the sympathetic author of this treatise used him and other diviners to belabor the hypocrisy of those who did.[8]

Whether slavery was part of the order of things or a matter of convention did not inspire philosophical discussion as it did in Athens. Whether it existed in China is debatable for the fifth through early third centuries B.C. An estimate for the Han period puts the proportion of slave to free at about 1 percent, with a large margin of uncertainty.[9] We do not hear of slave authors or slave physicians, as we do for Greece.

Origins of Scientists and Physicians

There is a good deal of concrete evidence on the social origins of cosmologists and scientists of various kinds. Much information from both cultures is anecdotal. It is difficult to be sure how representative many cases are. Biographies and stories suggest that in China, from the late Warring States era on, many of the literate elite had a common background, but in the Greek world the pattern was not at all uniform (see p. 89).

Philosophers and scientists generally were born into the shih, which gave them an opportunity to be educated (see p. 17). Only a few people found unconventional routes to literacy and even literary eminence. The most celebrated exception was Lü Pu-wei (d. ca. 237 B.C.), whose biography starts out, "He was a great merchant of Yang-ti, who traveled back and forth buying cheap and selling dear, until his family had laid by a thousand ounces of gold."[10] He used his fortune and his shrewdness to attract and groom a noble dependent and make him king of Ch'in. In time Lü's protégé fathered the First Emperor. Lü became chancellor of Ch'in and took the initiative in compiling the classic *Springs and Autumns of Master Lü*, which has much to say about science and medicine (p. 32).

The ju, a subset of the shih, were important in traditions of learning. Authors inside and outside the Confucian lineages used "ju," a word of unclear origin, to label followers of the sage. As often happened in China, thinkers looking for precursors read this label into the past. Historians still define ju as a class of teachers, to which Confucius belonged, who tutored descendants of noble families in manners, morals, or basic skills. In his time, however, the word did not refer to such a group. In the *Analects*, Confucius uses it once, ambiguously and evidently not referring to himself. His successor

Mencius, nearly two hundred years afterward, was the first to use "ju" for the stratum to which he himself belonged.[11] Even in the Han period, authors used it sometimes for members of a lineage that claimed descent from Confucius, but more often for conventional scholars regardless of affiliation. Initiated ju used it mainly to express their own solidarity, excluding those who did not belong in their ranks.[12] When the rejected used it, they implied snobbishness and pedantry. In the second century B.C., the state set up a school to teach prospective officials five widely studied classics that the ju who engineered this move claimed as their own. This gave the five—in the eyes of the state and eventually in the minds of aspirants to officialdom—a status above that of other texts.

Physicians in the Greek world set down their doctrines in writing as early as the philosophers did, but this was not the case in China. The status of doctors in Confucius's lifetime, about 500 B.C., was well below that of shih. Early sources (such as Narratives from the States) often speak of physicians as artisans, and artisans as hereditary servitors. One of the ritual anthologies describes an idealized antiquity: "Those who practiced an art in order to serve their superiors—exorcists, scribes, archers, charioteers, physicians, diviners, and miscellaneous artisans—did not follow a second occupation and did not move from one appointment to another. When they left their districts, they were not treated as equal to officials."[13] There is every reason to believe that throughout the period that interests us, the great majority of therapists inherited their occupations and were unable to read and write. More of them were practitioners of popular religion than of the evolving high therapeutic tradition, with its links to cosmology. Their skills were more often priestly than intellectual.[14] But there were elite physicians as well.

In early historical accounts of individual physicians, nearly all of those recorded were wellborn. Table 1 demonstrates that this continued to be the pattern for centuries after the end of our period. Of the seven late Chou and Han figures, their ranks or patients indicate

Table 1 *Social Backgrounds of Early Doctors*

DYNASTY	NAME	BACKGROUND
Legendary	I Yen	Physician to ruler
	I Huan	Physician to ruler
	I Ho	Physician to ruler
Chou	Wen Chih	Treated king
	I Chü (?)	Treated king
	Pien Ch'ueh	Hostel-keeper (legendary?)
Han	Chang Chi	Prefect
	Ch'un-yü I	Aide to Director of Imperial Granaries
	Lou Hu	Governor
	Kuo Yü	Hermit, beggar
Post-Han to 550	Hua T'o	None, but probably legendary
	Chang Tzu-hsin	Man of letters, Chief Steward of palace medicine
	Ch'u Ch'eng	Son-in-law of emperor, son of Imperial Secretary
	Ch'u Kai	Aristocrat, Imperial Physician, ennobled
	Hsu Chih-ts'ai	Official and courtier in three dynasties
	Hsu Wen-po	Great-grandson of Prefect
	Hsu Ssu-po	Cousin of Hsu Wen-po
	Hsu Yü	Doctor, Director of Palace Attendants
	Huang-fu Mi	Famous scholar, recluse
	Li Hsiu	Son of Adjutant to Heir Apparent
	Ma Ssu-ming	Supervisor of Pharmacy to Heir Apparent
	T'ao Hung-ching	Tutor to Princes, imperial favorite
	Wang Hsien	Descendant of Wei Minister of Works
	Wang Hsi	Imperial Physician
	Yao Seng-yuan	Hereditary doctor, Principal Physician, ennobled

that four were aristocrats. We cannot be sure that Pien Ch'ueh was more than a collection of legends. His position as a hostel-keeper is ambiguous, for sinecures of that kind ran in *shih* families.[15] Kuo Yü the hermit and beggar fits a common picture of the hidden sage. Since both were literate, it is unlikely that their backgrounds were lowly. Not a single person in this list is an unquestionable man of the people—obviously a matter of the historians' bias.

In any case, enough members of the literate, officeholding stratum were seriously involved in the practice of medicine to be responsible for the early medical classics. Why did they choose what most people considered a menial occupation? To answer this question, we would need to understand the diversity of health care before the Ch'in conquest better than we do. A couple of pieces of the puzzle are, however, reasonably clear. As part of the social changes from the eighth century B.C, on, downward mobility among the sons of aristocrats landed them in jobs that their ancestors would have considered beneath their station (p. 17). Some rulers and nobles of states large and small wanted to be treated by wellborn people. By the Han, in fact, some learned physicians were touring from court to court and doing very well at it.[16]

With astronomers the pattern is simpler. The authority to observe the sky and work out ways to compute the calendar was part of the imperial charisma and was naturally of great concern to rulers and to those who wanted to rule. The standard history of the Eastern Han specifies the cosmic duties of the Grand Scribe: "He is in charge of the seasons of heaven and the ephemeris of the sun, moon, and planets. When the year draws to a close, he submits a new calendar to the throne. For all state ceremonies of worship, funerals, and weddings, he is in charge of submitting a propitious date and time, along with forbidden times. Whenever the state receives propitious or baleful omens, he is in charge of recording them."[17]

These duties combined mathematical astronomy (li or li-fa) and astrology (t'ien-wen). The two fields were complementary.

Practitioners of the former designed the calendar (or, more accurately, the emphemeris) to predict a number of celestial phenomena, not just to count days, as modern ones do (see p. 227). Astrology comprised observing unpredictable phenomena (to spot ominous happenings in the sky and on the earth), recording and interpreting omens, keeping time, and divining lucky days. The court astronomers drew on meticulous records of phenomena that their predecessors had kept for centuries.

The post of Grand Scribe originated in the early Chou as that of chief ritualist and evolved into that of chief palace scribe, historian, diviner, and astronomer, subordinate to the Chamberlain for Ceremonials.[18] The biographical data on Ssu-ma T'an and his son Ssu-ma Ch'ien in the late second century B.C. suggest that they were descended from annalists, and do not include computational skills. When they wrote the first of the classic histories, *Records of the Grand Scribe*, they included in it the sky lore of their time; the next history in the series included the computational methods as well. Chang Heng in the second century A.D. was renowned as an expert in mathematics, astronomy, and cosmology and as the mechanician who invented the first water-powered armillary sphere (see Table 2). Of the twenty experts who spent thirty years arguing over and preparing for the calendar reform of 104 B.C., the standard biographical references ignore all but one, a recluse from the far west of China (a recluse is a gentleman eligible for office who, sometimes ostentatiously, refused a post). The three who planned the second important reform of the Han, that of A.D. 85, are equally obscure. The only subordinates of the Grand Scribes who clearly needed mathematical ability were officials of fairly low grade. Some of the great private astrologers, who taught disciples, took part in discussions or were coopted into the astronomical bureau.[19] It is likely but not certain that all the astronomers outside the bureau as well as those in it were highly educated members of the elite. Not all those in the palace, especially those in the higher ranks, needed to be skilled in computation.

Mathematics is more difficult to pin down. It was not an occupation but a set of skills. Some of the elementary techniques were probably part of every gentleman's education. Others were required for such duties as surveying, accounting, and planning, for which many officials were responsible at some time in their careers and which businessmen and craftsmen mastered as well.[20] Some extremely complex techniques made possible the official functions of astronomy and mathematical harmonics (the study of relations among musical sounds). The latter was as important in China, because of its connections with the music of court ritual, as in Greece, where it had no such links.

We can conclude, then, that the abilities underlying science and medicine were diffused through many levels of Chinese society but were concentrated near the top because of the elite's literate foundations. This social dispensation was shaped to some extent by patronage before the Han and official status during it.

Employment and Patronage

Throughout our six centuries, in independent states and then in the imperial court, governments employed a good many literate functionaries. Some were highly cultured, and others (especially before the first century B.C.) were barely able to read and write. The most usual title for learned scholars, Erudite, was originally a label for broadly learned ritual and political consultants, who until the mid-third century B.C. were not regular officials and until the late second century B.C. generally had no teaching duties.[21] Titles that smack of learning, such as Tutor to the Heir Apparent, were often honors granted to eminent officials and do not imply special qualifications to teach. Patronage supported a few eminent intellectuals before the Han, and employment gradually supplanted it afterward.

"Patronage" can mean a great many things. In historical writing, the most obvious meaning is "substantial individual support without secure employment." That is what it meant in Europe in the sixteenth and seventeenth centuries, when dynasts great and small supported scientific celebrities and underwrote the publication of treatises by potential stars.[22] In China patronage had no statutory standing. The sovereign could also terminate the appointments of regular civil servants when he wished, but that was rare. In practice, infractions or imperial tantrums were more likely to lead to a fine, exile, confinement, or an order to commit suicide. Clients, on the other hand, were dispensable on any whim. We will examine in the next chapter the complex forms that patronage took in Greece and Rome (p. 96).

In China certain beneficiaries of patronage were the so-called guests (k'o) at local courts in the Warring States period.[23] Patronage first became prominent a little before 400 B.C. Its vogue lasted a couple of centuries, until the new imperial government wiped out the old local courts, leaving few large-scale patrons.

Only a handful of the clients that patrons supported were inquirers after wisdom. Guests were more likely to be masters of useful arts, advisors on rulership and strategy, trainers in military techniques, persuaders, confidential messengers, assassins, or experts in dirty tricks. Among the unlikely clients who turned out to be indispensable in emergencies were an expert in shouting loudly, a man whose only special skill was crowing like a rooster, and a burglar who pursued his calling in the guise of a dog.[24] Patrons strove to attract guests from the courts of rivals, knowing that they would bring current intelligence with them. Some conventional embroidery aside, there is no reason to doubt the consistent reports that the most avid and wealthy patrons supported assorted clients by the thousand. Large-scale patronage persisted, as we shall see (p. 33), in the lifetimes of the historians who described its earlier phases.

Wisdom was desirable in such guests, but no more so than a quick wit or skill at deception. Anachronism has magnified the importance of philosophy in the feudal courts. In the late fifth century B.C., science and medicine were recorded in philosophical writing or not at all. Patrons wanted neither basic research nor innovative perceptions, but advice and other services that would help their states survive if they were weak or grow if they were large. If someone turned up offering new military technology or fresh uses for scurrilous rumors, that was all to the good. Benefactors did not begin with want lists when seeking their proverbial three thousand guests.

Grants from the ruler's purse or sometimes honorary appointments, revocable without notice, supported clients. Retainers who had no official status were competing with high civil servants— whom they had no standing to challenge—as well as each other. Regular officials and other courtiers considered their own advice adequate, after all, and were annoyed when their superiors' pets interfered.[25] Individual ability to keep the favor of a ruler was as important to Chinese clients as open debate was for philosophers in Greece (see p. 122). Sometimes guests survived because they had skills that civil servants lacked or were too conventional to use, and sometimes because they offered an alternative to advice from officials that the ruler did not want.

Shortly after the T'ien family of Ch'i usurped the power of the original ruling clan, they began to collect, among other recipients of their conspicuous patronage, a few diverse philosophers. Duke Huan of Ch'i (r. 374–357 B.C.), who became the paramount ruler in the northeast, began (if we credit the legend) supporting a number of clients, but his successor, King Wei (r. 356–320 B.C.), did not try to outdo him. The ruler after that did, and was fastidious in his choices. According to an account in *Records of the Grand Scribe*, "King Hsuan (r. 319–301 B.C.) was fond of literary studies and of gentleman-retainers who were traveling advisors. Among his seventy-six [guests] were Tsou Yen, Ch'un-yü K'un, . . . and their like. He granted them all mansions and made them Senior Grand Masters. They did not govern but did take part in policy deliberations. As a result, 'the scholarly gentleman-retainers of Chi-hsia' flourished again, amounting eventually to several hundred and then several thousand."[26]

The long passage of which this is part lists only half a dozen illustrious philosophers out of the seventy-six guests. The king housed these dependents in the neighborhood of the capital's Chi Gate. This and other accounts make it clear that he valued them not for their reflections on ethical or cosmological topics but rather for their aid in Ch'i's intrigues with and against its neighbors. The early sources mention only guests, never formal organizations, never institutions or any buildings but residences.

Modern historians of philosophy have wishfully invented a Chi-hsia Academy, "a meeting place for intellectuals [that] can in some respects be likened to some of the larger government funded research centres today."[27] But the main item of evidence offered for this fantasy, the title Senior Grand Master in the passage just quoted, was an honor with no institutional status attached. There was no Library of Alexandria, much less a Rand Corporation, in Old Cathay.

The kings of Ch'i were still putting up appreciable numbers of guests in the quarter near the Chi Gate half a century later, when the celebrated Hsun-tzu was there under King Hsiang (r. 284–265 B.C.). Thinkers as important as Mencius, Kuan-tzu, Sun-tzu, and perhaps Chuang-tzu accepted the hospitality of the Ch'i kings in one connection or another.

What did patrons hope for from these few philosophers, in addition to what the crowd of pragmatic experts offered? Anecdotes and surviving writings of clients point to guidance for what nowadays would be called order and control. Rulers wanted the secrets of effective rule. They wanted rational solutions to problems of policy and administration—to impose order on disorder. They wanted justifications of government that would build support for the state. And, not least, they wanted widely admired people at their beck and call, ready to entertain their courtiers and confound their enemies.

From the viewpoint of the philosophers, especially but not only Confucius's posterity, the world was falling apart. They admitted that keeping it together depended on the whims of kings and princelings. Persuading the rulers to discipline themselves, to avoid making life impossible for their subjects, and to stop interfering with the work of their officials was as much as they could hope for. Neither side could win this tug-of-war, but the scholars sometimes persuaded the rulers that they had long-term interests in common.

Although patronage became a fashion, the taste for scholarly clients did not spread with it. When T'ien Wen, lord of Meng-ch'ang, succeeded his father in 298 B.C., he "sought out the guests of the feudal lords, even fugitives guilty of crimes, all of whom went to him. The lord of Meng-ch'ang used his patrimony to entertain them generously, so that he was able to compete for guests from all over the realm. He fed several thousand guests, treating them all as he was treated." The biography of which this is part does not describe a lover of philosophy. It shows a lord taking pains to impress a motley assortment of guests who ranked far below him.[28]

In the early third century B.C., in the states that were competing to rule all of China, great patrons were building collections of clever, dangerous, or otherwise useful people. This was a war of hospitality. The prominent exception was Ch'in—the eventual winner—whose rulers had no such collector's instinct. Whether failing to collect guests was wise became problematic when three of the great regional patrons combined forces in 258 B.C. to deal Ch'in its first important military defeat, delaying its unification of China by a generation.[29]

This defeat was responsible for the momentous collection of guests that Lü Pu-wei built up when he ran the state of Ch'in at mid-century. As chancellor, Lü was all too aware of the three patrons who had frustrated Ch'in not long before his time, "lowering themselves to associate with gentleman-retainers and taking pleasure in their guests, one outspending the other. He felt that, in view of Ch'in's power, it was embarrassing not to measure up to them. He summoned and attracted gentleman-retainers, treating them generously, until he was feeding three thousand guests. At this time in [the courts of] the feudal lords there were a number of masters of argumentation, such as Hsun-tzu and his like, writing books that spread throughout the realm. Lü had each of his guests set down what he knew about."

He edited a selection of these discourses, or had them edited, into the eponymous *Springs and Autumns of Master Lü* (ca. 239 B.C.). This was the first of three important compilations that put monarchy and morality on a cosmological foundation. Its sociopolitical goals were not at all ambiguous. Lü aimed his program at the conquest and unification of the known world by his young ruler, who later became the First Emperor.[30] Lü was the patron, not the author, of this large, influential scholarly enterprise, but it would be naive to deny that he defined its focus and imposed a good measure of unity on it. That suggests exceptional acuity, in addition to the political power that put some of the best retainers of the time at his disposal.

Two details of the quotation from this work are germane to the question of patronage. First, when the text speaks of some guests as "masters of argumentation (*pien-shih*)," it characterizes their fame by books, not face-to-face discussion—a reminder that in China writing was the typical form of dispute. Second, *Springs and Autumns of Master Lü* is obviously not the collective work of three thousand hands. There is no reason to imagine that philosophers were more than a minuscule fraction of Lü's guests. Only three or four are needed to account for textual inconsistencies, and a dozen would have been more than enough to provide the many viewpoints on the microcosm and other issues in this generally coherent book (see p. 212).[31]

After another century, by around 150 B.C., the Han imperial regime was settling into place. Only a handful of princely courts that could support clients survived. These were the fiefs that the founding emperor of the Han had awarded to relatives. Eventually, as the old aristocratic ways spread slowly down the social scale, mere officials and even merchants collected guests. Still, the only evident support for scholars who discussed the cosmos was near the top. Perhaps only those with great power or wealth were likely to take seriously claims about the practical uses of philosophy. As always, it competed for their largess with many other enthusiasms.

The last of the philosophical projects held together by lordly patronage, rich in discussions of the phenomenal world, was that of Liu An, king of Huai-nan (ca. 180–122 B.C.). Liu indulged in patronage and literary synthesis on a scale reminiscent of Lü Pu-wei. His biography records that "he attracted to himself several thousand guests and gentlemen who practiced technical arts."[32] Anecdotes about the surviving *Book of the King of Huai-nan* claim that a mere eight of these retainers wrote it.

Liu compiled it to present to his newly crowned nephew as a guide to rulership. The Martial Emperor did not accept its guidance and perhaps came to view his uncle as competing with him for prestige by shows of patronly magnanimity. The king of Huai-nan and his retinue, philosophical and otherwise, were anachronisms. Since 180 the central authorities had been fitfully eliminating the local kings, who interfered with centralization and were conceivable rivals for the throne. The government claimed that its military expeditions against them were responses to rebellions, but some of these it imagined or provoked as occasions to consolidate its own control.[33] After the government drove the king of Huai-nan to suicide and had his household slaughtered, no great local patron was left to collect even a handful of philosophers with technical interests.

By Liu's lifetime, outside such niches as his court, employment in a highly ramified bureaucracy had long since replaced patronage as a living for intellectuals. The end of the local rulers left a void for imperial patronage to fill. The Martial Emperor—like the First Emperor of Ch'in, with whom he had more than one affinity— became notorious for the favors he showed, and the wealth and power he granted, to a few occultists.[34] But by Warring States standards neither emperor would have been taken seriously as a collector of experts.

This was the same Martial Emperor who allowed only the classics favored by ju to be taught by the state. In supporting opposed interests he was no anomaly, however. Monarchs who considered themselves above the laws eventually tired of ministers who kept protesting their disregard of ritual codes and their flouting of precedents. Guests and eunuchs indulged emperors' fantasies that civil servants were in duty bound to frustrate. Favorites could flourish, or indeed survive, only in niches that those fantasies created. Despite the fury of officialdom at being bypassed and the managerial catastrophes that were inevitable when a ruler handed over authority to unaccountable favorites, the dream of untrammeled power led some monarchs to prefer clients over civil servants as long as the empire lasted.[35]

The importance of monarchic patronage faded in the Ch'in and Han periods as the new order expanded the civil service that local courts had pioneered. The Han state imagined and worked out ideal, rational, invariant structures of government. If a bureaucracy is a formal structure that endures, and aims to function identically, regardless of what individuals fill its posts, a bureaucracy is what the Han instituted.[36] Appointment still depended heavily on birth, but the government did take a large step toward the mature system of a thousand years later.

When discussing the relations of officials to the emperor, we cannot assume that the former were a uniform group. The very highest officials were noble intimates of the ruler and routinely spoke for him. He could not reject their advice out of hand. Often it was they rather than he whom scholars, who generally held appointments in the lower ranks, had to persuade. Thus "the emperor" can be shorthand for "the emperor, the Three Dukes, the Grand Masters, and the rest of the high officials."

The middle and lower ranks of the civil service provided secure careers for large numbers of officials and regularly monitored their work. The government mainly needed functionaries trained to carry out routine court functions and experts on precedent who could maintain established usages. Those responsible for significantly technical tasks were a decided minority. Although the bureaucracy increasingly expected recruits to be familiar with conventional classics and to speak and write well, most positions required little rational analysis and no original thought.

The most prominent positions for people with technical skills were the offices of the Grand Scribe (in charge of astronomy and astrology) and the Imperial Physician, both civil servants of fairly low rank. The former not only supervised experts but operated an observatory of impressive dimensions; the site of one built between A.D. 56 and 59 has been excavated and studied.[37] There were two Imperial Physicians in different parts of the court, each with specialist aides, and others in the entourages of nobles and other dignitaries.[38] Other agencies of the Han civil service provided almost all the technical functions the court needed: the offices for divination, music, and harmonics, various others under the Chamberlain for Ceremonials, and still others in one bureau or another. These specialized organs employed many technicians in low-ranking, mostly hereditary positions. None but the most ambitious could rise far.

The centralization of power was fateful in several ways. Once it was substantially complete, only the palace could afford new technical activities on a large scale. Inevitably, that is where such elaborate mechanisms as the mechanical seismograph of A.D. 132 were invented and perfected. With the competition for prestige among rulers a thing of the past by 100 B.C. (p. 32), there remained only one emperor, whose fancy one could hope to strike. Innovations that did not appeal to him were generally abortive.[39]

A true bureaucracy, aimed at routine above all else, does not demand innovators or iconoclasts. For the Han we know the names of only a few enterprising technical officials. A handful of shih, even commoners, could do well enough outside the system if they could fascinate the ruler. Few Han rulers developed a taste for anything more mentally taxing than the pursuit of immortality. A majority of them were indifferent or hostile to classicists, and only a few considered Erudites of any use in forming state policies.[40] They were willing to appoint people to perpetuate old traditions, not to launch new ones. Given the general aspiration toward civil service positions among members of the elite, the highest levels of society tended to accept this bias. The shift in classical studies toward narrow scholasticism in the first and second centuries A.D. (p. 56) was part of this conservative trend.

The careers of a few important contributors to cosmology reveal some general trends. We summarize them in Table 2. The first three scholars listed in the table were the authors or sponsors of the three great synthetic works that largely define early Han cosmology and that give us substantial discussions of themes in science and medicine before these fields developed their own classics (see p. 261). Lü's *Springs and Autumns* and Liu's *Book of the King of Huai-nan* are collaborative works, privately commissioned by powerful patrons. The writings of Tung, an official of the mid-second century B.C., had some influence on imperial ideology.[41]

Of the remaining nine authors in the table, only three, Liu Hsiang, Liu Hsin, and Chang Heng, held high regular civil service posts. The Lius, imperial librarians, assembled a large portion of the pre-Han classics from manuscripts (many of them disordered) that they found in scattered government collections and cajoled from officials.[42] Liu Hsin's contributions to natural philosophy continued his father's interests in astrology and divination. Chang Heng was an innovative

Table 2 *Careers of Important Philosophers*

NAME	DATES	HIGHEST POSITIONS	IMPORTANCE
Lü Pu-wei	d. 237? B.C.	Chancellor	Synthesis
Liu An	179?–122 B.C.	King	Synthesis
Tung Chung-shu	179–104 B.C.	Erudite; Administrator of kingdom	Synthesis
Meng Hsi	fl. 69 B.C.	Office Chief	Divination, *Changes*
Ching Fang	77–37 B.C.	Court attendant	Harmonics, *Changes*
Liu Hsiang	79–8 B.C.	Chamberlain for the Imperial Clan	Restoration of classics, divination, astrology
Li Hsun	d. 5 B.C.	Commander-in-chief, Cavalry	Astrology, divination
Liu Hsin	d. A.D. 23	Superior Grand Master of the Palace	Restoration of classics, bibliography
Yang Hsiung	53 B.C.–A.D. 18	Court poet	Cosmology, literature, Confucian doctrine
Huan T'an	43 B.C.–A.D. 28	Court Gentleman for Consultation	Philosophy, criticism of reliance on portents
Wang Ch'ung	A.D. 27–97	Prefectural clerk, then recluse	Skeptical but not rationalist philosophy
Chang Heng	A.D. 78–139	Imperial Secretary, Grand Astrologer	Cosmology, astronomy, invention, poetry

polymath who made important contributions to mathematical astronomy and mechanical invention as well as poetry. He would have been exceptional in any period.

The remainder, interesting though their work was, occupied sinecures and largely spent their careers as imperial advisors. Military titles merely reflect the fact that civilians controlled the bureaucracy of war and were regularly appointed to commands. Such titles no more describe a career than do such generic categories as "court attendant" or "court gentleman for consultation." Yang Hsiung was a dismally unsuccessful courtier. Only after his lifetime was he recognized as the leading cosmological author of his time. Wang Ch'ung never qualified for a palace appointment; his celebrated *Discourses Weighed in the Balance*, written in reclusion while a private teacher, seethes with frustration. Although the civil service indeed supported major figures in cosmology, it did not on the whole support them as cosmologists. The same can be said of most leading practitioners in other technical disciplines.

The writings of the anonymous intellectuals from the first century B.C. on who founded predominantly technical traditions reflect the state's control over their fields.[43] Early scientific writings show government influence, which varies in type and strength from one field to another. It was strongest in astronomy, which was essential to the operations of the state, and weakest in alchemy, which could count on no central support except when a practitioner convinced an emperor that he could make him immortal. It is productive to divide the sciences into the subsidized (that is, regularly supported as part of the civil service) and the unsubsidized.

The subsidized sciences paid a price in government control. In the Han, the state's uses of mathematical astronomy shaped it. The calendar was an item of imperial regalia, ceremonially granted at the new year to everyone in the realm. Most of the precise timekeeping and predictions of celestial phenomena in it were of no use to the population at large. It was particularly useless to farmers living in a wide range of climates, although the calendar was supposed to regulate their work everywhere. It was the symbolic and ritual significance of the calendar to the palace that mandated ever-increasing accuracy and kept astronomical officials endlessly revising the computational techniques. Astrology affected only the dynastic house. It was inapplicable to the experience of others, who could draw on many non-astrological techniques of divination.[44]

Subvention and regulation by the state tended to ensure the persistence of written traditions. The basic documents of Han state astronomy have survived intact. From the official histories we can even reconstruct to some extent private studies that the government did not support or encourage but from which its astronomy often drew vitality.

Of writings on alchemy in the Han, all we have today is a single short text. It appears to spring from a private lineage of the kind that was responsible for compiling the classics of other sciences and medicine (p. 60). As for siting (or geomancy, the science of placing houses and tombs in the landscape), apparently no extant book comes from the Han, although it is possible that some existed by that time.[45] Later traditions trace both arts to classics supposedly by the Yellow Emperor, whose legend in the mid-Han often makes him a revelator of technical knowledge. We have not so much as a word about any Han alchemist or master of siting who is more than a legend.

Medicine and mathematics do not fit neatly in a framework based on subsidy. People up and down the social scale practiced both (pp. 23, 27). The *Mathematical Methods in Nine Chapters* (ca. A.D. 100) is a collection of numerical problems of many kinds (see p. 230). They resemble those treated in the writings on arithmetic and mensuration in the Euclidean Corpus, in the books of Hero of Alexandria, and in those of Diophantus (in Greece, a tradition lower in prestige than geometry). The Chinese treatise, unlike most early Greek writings, was not meant to be adequate by itself.

It presupposes a teacher who will explain how to apply each problem to pertinent instances and how to get from the laconic statement of one solution to a method for dealing with a class of problems. Although it began as a textbook, it soon filled the need for a classic.[46]

Medical doctrine is even more difficult to trace to its origins. The earliest classics transmitted to the present, the *Inner Canon of the Yellow Emperor* (probably first century B.C.) and its immediate successors, take the form of dialogues between an emperor and his ministers. Some earlier writings have been excavated, but they are therapeutically oriented. Several reflect a doctrinal foundation of popular religion and occult belief rather than secular rationalization.[47] There is no evidence in the Yellow Emperor classics that they were written by officials or especially for governmental use. The earliest collection of materia medica (p. 232), incorporates the language of both political hierarchy and belief in immortality, ruling out a certain judgment on its origins.

The symbology of the emperor as mediator affected mathematics, alchemy, medicine, and materia medica as strongly as, though in a different fashion than, it did astronomy. For that matter, it was no less potent in poetry connected with the court.[48] In the classics of the technical traditions, dialogue forms and discussions of macrocosm and microcosms derived from the state-supported classics (p. 59). This was more than mere decoration (except in the case of alchemy). The official canon and the Ch'in-Han synthetic compilations imbued the early technical classics with a portion of their meaning that contemporary readers took quite seriously.

We can draw several conclusions from this evidence. Patronage before the Han did not give philosophy a place of any particular importance. Nevertheless, this informal institution shaped philosophical ends and means. Confucius's writings reflect his frustrated quest for the role of official advisor before patronage appealed to rulers as a source of prestige. Two generations later, it had attracted a few. By the early part of the third century, any thinker who could compete with tacticians and intriguers might win a chancy livelihood. The few intellectuals who found their way to the courts of Ch'i and other states were a wildly assorted lot, and they developed philosophy in a great many directions.

The unification of China changed all that. The new imperial pattern was not defined or imposed overnight, but within a century, as we have seen in the case of the king of Huai-nan, no room was left for the old diversity. The concerns and interests of the state naturally favored a narrow ideological range, which most philosophers of the time accepted. Even those who opted out of the system largely responded in its terms.

Those who wrote about the sciences from the first century B.C. on created distinct universes of practice and meaning. But in doing so, they did not banish from their writing the political ideals that had shaped the Han empire. There is no reason to suppose that they were under overt pressure to incorporate those ideals. Scientific and medical authors, who came from the tiny literate minority, conformed to conventional values and accepted the priorities of the state.

The state succeeded in shaping the aspirations of gentlemen, those who practiced the sciences and medicine among them. But below these strata of society were very large numbers of therapists, people who solved mathematical problems, and intelligent (if seldom fully literate) craftsmen whose work the state did not influence. They generally passed down their technical knowledge and skills to their children or apprentices without any formal schooling. Although not many uneducated people could support themselves in surveying or other mathematical arts, others besides sons or pupils of healers made a living at therapy. Diviners and astrologers as well as priests of the popular religion dealt with therapeutic matters.[49] Although we do not know enough about technical skills among commoners to draw more general conclusions than these, archaeological excavations may throw additional light on this topic, as they have on many others.

Individuals, Groups, Education, Transmission

There are large and important differences between what held thinkers together in China and Greece. Here we are mainly concerned with what we might loosely call Chinese institutions for higher education and collective research. How common was it for philosophers, scientists, and physicians to belong to any sort of collectivity? What did membership entail? How were they organized? How were members recruited, what constraints did they accept, and how did they view deviation or defection?

Individuality—personal character, original points of view, iconoclasm, and idiosyncrasy—was not at all rare in early China. Every philosopher down to 250 B.C., and most of those later, spoke with a characteristic voice about a personal vision. In fact, the visions of a few early intellectuals, especially Confucius, Mencius, Hsun-tzu, and Mo-tzu, decisively ruled out a conventional civil service career.[50] Even those who wanted to conform were critical about what one should conform to. They were, in other words, no less complicated than Greeks, or us.

Hsun-tzu, the most aggressive of Confucius's successors, although he ostentatiously insisted that his readers should refuse office under any ruler who lacked virtue and public spirit, was one of the most conspicuous beneficiaries of patronage. The king of Ch'i whose generosity he eagerly accepted was no exemplar. The desire of the royal clan to overcome a reputation as usurpers (see p. 29) seems to have motivated their taste for collecting guests.

When writings on the sciences appeared, some were handbooks without personal flavor, and others took the form of master-pupil dialogues that paid little or no attention to characterization. In both cases that is a matter of genre. But Yang Hsiung's *Supreme Mystery*, a landmark of Han cosmology, managed to combine systematic analysis, cosmological depth, high literary elegance in both form and content, and sheer quirkiness.

If we use "individuality" less broadly, for thinkers who refuse to identify themselves with any group, we find few of them. As soon as one of those few became influential, a lineage tended to grow out of his teachings and to aim for their permanent transmission.

The quintessential Chinese individualist was Chuang-tzu, for whom social norms were mere impediments to a full life, group solidarity was misdirected, and virtue a matter for ridicule. We know nothing about him but legends. Not long after someone set down his paradoxical anecdotes, one or more lineages were transmitting this initial group of writings and, in their own teachings, turning his ideas in new directions. When one such group compiled the book named for him at some time between the late third century and 100 B.C., it put together writings by at least five people or groups at various dates, encompassing different agendas but aping the anecdotal and paradoxical style of the earliest Chuang-tzu.[51] That was his fate: to inspire stylistic imitators. The later authors added many splendid anecdotes and several new themes but made no further breakthroughs in iconoclasm.

Once the Han age of unity was under way, debunking ceased to be a major activity of philosophers. It revived as the Han order disintegrated, in a succession of individuals who belabored the credulity and narrowness of the scholarly rank and file. The best-known three are Huan T'an (43 B.C.–A.D. 28), a music official of low status; Wang Ch'ung, the frustrated author of *Discourses Weighed in the Balance*; and Wang Su (195–256), who became a high functionary under the successor dynasty to the Han. Huan T'an inspired the other two when he denounced the ascendancy of scholasticism (see below, p. 56), and the exaggerated belief in divination and in Confucius as a preternatural savior figure that was spreading from the first century B.C. on.[52] The careers of Huan and Wang Ch'ung suffered for their rejection of the conventional wisdom. What stands out is their commitment to an abiding tradition of dissent in the name of Confucius versus the self-styled Confucian orthodoxy that they strove to beat back. They were nonconformists, but hardly lone adventurers of the spirit.

The Chinese norms, then, were identification with a group and aspiration toward an imagined orthodoxy (although, as in Huan's case, from time to time norms inspired protest). They were the mirror image of the Hellenic emphasis on a thinker's own ideas even when he belonged nominally to a group.

The history of education in China, also unlike the Greek case, was largely a history of **collectivities**. Those who worked out influential principles of education—for example, Mencius and Hsun-tzu—were not encouraging displays of individual talent but were largely concerned with personal cultivation and the social utility of ancient wisdom.

The coterie of master and disciples was the basic unit of education at every level. Philosophers from Confucius on almost always taught adult or nearly adult disciples, in numbers that reflected the teachers' reputations. The devotion of pupils generation after generation, in a relationship modeled on blood kinship, generated much of that fame.[53] Social and official status were also components, fluctuating in importance over time. In some cases the decision to become a pupil was that of the adult individual, and in others, of his family. Payment was the norm, but it is seldom possible to be sure whether a given teacher expected payment, or what the amount was, and usually hard to say whether he depended on it for a living.

Scattered through the extant records are a dazzling variety of coteries who left the merest traces by accident. This multiplicity reflects the wide range of thought about society, politics, and cosmology in the late Warring States era. A typical example is Hsu Hsing, about whom we know little more than that "he practiced the teachings of the Divine Husbandman." This, we are told, inspired him and his pupils to weave sandals and sitting mats for a living and led him to criticize his ruler for not doing his own plowing and cooking. His name survives mainly because the eminent Mencius troubled himself to harangue a disciple of Hsu's on the division of labor (see p. 16).

Few such groups outlasted the first generation of disciples. The self-designated successors of Confucius are unique because they maintained, manipulated, and documented their lineages over nearly twenty-five hundred years—with a fair bit of vagueness and some breaks in the record, to be sure. The Mohists—the followers of Mo-tzu—apparently maintained themselves for two or three centuries (see p. 55).

In considering what social relationships underlay education it will be well to ponder some advice from a chapter entitled "Respect for the Teacher," obviously meant for the adult student, in *Springs and Autumns of Master Lü*: "In studying it is essential to progress in learning in such a way that there will be no confusion in the mind. Memorize [the texts] avidly. Respectfully wait for a break in [what the teacher is saying], and if you see that he is in a good mood, ask about the meaning of the book. Make your ears and mouth obedient so that you do not contradict his intentions. When you have left him, ponder what he has said." This ideal of learning, although it stresses person-to-person teaching, is centered on a written book. The key is the pupil's receptivity as the teacher expounds the text. The notion is authoritarian, but it anticipates the teacher's solicitude in response to the disciple's devotion and obedience.[54]

The relationship was not necessarily exclusive. The analogy with blood kinship implies that an individual will have only one mentor, and some authors portray such relations as inherently lifelong. Nevertheless, in practice, becoming the disciple of more than one master was neither blameworthy nor rare in either philosophy or science. We have records of such relationships, and differing versions of the same text excavated from the same tomb imply them. Both types of evidence indicate that the motive to take on another master was generally initiation into more classics or more versions of a given classic than a single teacher offered.[55]

Because formal education began with memorization, the classics embedded themselves in the student's consciousness and molded his growth. Just as important as the archaic words were the contemporary meanings that teachers poured into their explications in the name of fidelity to the past. The idea of virtue as emulation of antiquity encouraged among the elite a common language and an evolving set of shared values for coping with a constantly changing world.

The aftermath of the Ch'in unification transformed scholarship. The First Emperor tried to destroy proscribed books in private hands, and a great many more disappeared in civil wars.[56] Some texts survived only because someone had a copy hidden away or preserved a text in memory. The attrition left a permanent imprint on intellectuals, especially ju. From then on, many of them saw a primary object of education to be ensuring that the classics of their tradition were not lost. That made memorization of the exact text and meticulous copying of manuscripts all the more important. The same impulse also eventually persuaded the Han government to guarantee the survival of classics by appointing scholars to the court as Erudites.

The lowborn founder of the Han dynasty originally despised eggheads, but he acquiesced in recruiting officials who, among other specifications, could draw on "the techniques of prior sages." Representatives of the ju lineages were only one set of lobbyists among many. Many of the Erudites were experts on classics of various traditions, but others were ritualists, not particularly oriented toward texts. The Martial Emperor, in 136 B.C., got rid of the consultants on all but five of the classics—the *Book of Documents*, *Book of Songs*, *Book of Changes*, *Spring and Autumn Annals*, and *Book of Rites*. These five had been widely used for some time, but masters who claimed links to Confucius's teachings were promoting them as a basis for orthodoxy. They now became a single canon of scriptures to undergird a reform of ritual and education and to serve the larger goal of a state centered on the emperor's person.[57] By ending support for other canonic writings, this commitment persuaded the landowning, scholarly, officeholding elite to begin adopting the five as their own body of learning—one that, as interpreted by Han classicists, encouraged devotion to the state.

In 124 B.C. the government appointed quotas of official students to a Grand Academy and made Erudites responsible for educating these disciples in the rites and language needed to become officials and for passing down the classics intact to them. The Grand Academy became a permanent institution, but most Erudites served there only until promoted to an administrative post. There is no record that they shared activities, as the members of Aristotle's Lyceum did, or even met regularly. What they held in common was a bureaucratic title and the duty of producing graduates qualified (by literacy, not philosophical sophistication) for official appointment. Although palace documents ordered the creation of schools in prefectures, counties, and towns, historians have found more complaints about their absence than proof that officials built them.[58]

Monopolizing the Grand Academy was a mixed victory for the Confucians who made the subsidized classics their own. The government's interest did not lie in preserving the spirit of the sage's philosophy alongside the letter of the texts he had revered. In fact, the edict of 136 B.C. discarded, among others, Confucius's *Analects*, one of the texts that, up to that time, had had an Erudite to represent it. The new doctrines of monarchy drew important themes from most other philosophies of the time.

These promiscuous ideologies, which functionaries gradually learned to draw on, tacked decisively away from Confucius's teachings, although spokesmen for the state kept claiming that the political order was based foursquare on them. Let us look at a few examples. Confucius was concerned with the self-cultivation that would prepare a gentleman to advise a ruler without himself being corrupted by power. Han ideologists were preoccupied with defining separate spheres of authority for the emperor and his bureaucracy. Confucius, who did not often use the word "sage" (sheng), meant by it the ideally cultivated gentleman (or, once, an untypical ruler). His pre-Han successors also employed it for certain perfect rulers of the golden past, as well as for exemplary advisors. Spokesmen for the Han state applied the word mechanically to emperors— above all, to the current one. Confucius saw statutes and ordinances as a sign of failure to use moral example effectively. The state apparatus did not reject this idea, but its regular answer to problems of order was statutes and ordinances. Confucius was a humanist in the sense that he believed the problems of a good community could be solved entirely within the sphere of society. The state made concord between the cosmic order and the political order the key to social harmony.

We will speak occasionally of these new, reasoned beliefs in conventional elite thought as an orthodoxy, but the word needs to be used with care. Some early philosophers, such as Mencius and Hsun-tzu, the two most important successors of Confucius, emphatically denied that they shared ground with others. This does not mean that they favored every thinker's taking a unique stand. To the contrary, they excoriated rivals for impeding the orthodoxy that they believed was mandatory for a stable society (pp. 52, 64). The three Han synthetic works we examine—the *Springs and Autumns of Master Lü*, the *Book of the King of Huai-nan*, and *Abundant Dew on the Spring and Autumn Annals*—though seldom quarrelsome, gave a great deal of attention to what people ought to believe. Unlike earlier philosophy, they were, though far from unanimous, consistent in many ways, particularly on the cosmic and microcosmic foundations of the state. Ambiguity facilitated their consistency. Thus, despite the consensus that orthodoxy was desirable, a single, authoritative version of it did not come into existence. The word "orthodoxy" is mainly useful in connection with the government's shifting sponsorship of classics or scholarly lineages and its persecution of political activity or thought for being heterodox.

Early ideas of education did not formally divide the elementary from the advanced. Basic education aimed, not at all coyly, to maintain the status of elite families and to make their offspring useful to the state. No one expressed a desire to enable everyone to read and write; indeed, a high level of literacy was not widespread until after 1950. Gentlemen saw the social order as created and maintained at the top; whether the lower orders could keep records or sign their names was not a matter for concern. Some of the latter, even without formal schooling, learned enough reading and writing to master crafts or even to support themselves in clerical jobs below the official ladder. Government sponsorship of the classics encouraged the elite to reject any utilitarian standard of literacy that demanded less than their own massive ingestion of the cultural heritage.[59]

As for **higher education**, among the aristocrats of the Warring States era there was no fixed curriculum. Confucius set in train a new approach to preparing gentlemen, some of them marginal in social standing, for a life of public service. What mattered to him and his followers was the exemplary force of virtuous behavior shaped by ritual and moral self-scrutiny. Individual character was to be molded by the highest achievements of human civilization, which lay securely in the past. Although at first ju stressed ritual and personal cultivation, they gradually gave weight to study of the classics.[60] These books, surviving from antiquity, made the words of the sages available to the present. By the first century B.C., studying meant memorizing classics of great length.

As the state exerted authority over the preservation of texts and the training of civil servants, it did not invent anything like the later empirewide recruiting examinations for commoners. Representatives of the state from time to time tested the qualifications of candidates for the Grand Academy in a variety of ways. Officials informally quizzed them about their knowledge of policy issues, investigated their performance of previous official duties, confirmed that they were trained in a recognized tradition of a classic, or tested their ability to write official documents. But acceptance did not depend primarily on intelligence or skill. Here are the official qualifications, from the founding document of the Grand Academy: "those of seventeen years of age or older, of serious manner and deportment . . . fond of cultivation through study, respectful toward elders and superiors, with a respectful attitude toward the government's enactments and its moral teachings, compliant in their native places, not contrary in their goings and comings."[61] These attitudes and modes of conformity were preparation for the education of a bureaucrat, not of an innovator. An occasional but paramount responsibility of the Grand Academy was to educate future emperors; this mission was bound to influence the rest.

Records of its pedagogy have not survived. We do know that the quota of students formally enrolled grew from fifty in 124 B.C. to thirty thousand in the decades after A.D. 125. That the number of Erudites did not increase proportionately suggests, along with much other evidence, that these quotas were not filled with resident students. In fact, in the first and second centuries A.D. the fortunes of the Academy fluctuated between abandonment, forced enrollment of sons of officials, and the arrest of students by the thousand during confrontations between the emperor's powerful eunuchs and regular civil servants. Amid this turmoil, the classics taught varied considerably as scholarly factions fought quietly for status.[62]

As for private higher education, in the Eastern Han period many individual scholars accepted disciples. Disciples might follow more than one teacher or even more than one lineage (p. 59). Some masters attracted large numbers of pupils, depending mainly on the teacher's reputation not only for learning but for virtue and official status. Still, coteries that depended on political clout as much as on intellect were not likely to remain intact for many generations.

The stipends paid to teachers were traditionally low and (to banish all taint of the commercial) were not openly discussed. Their remaining low would explain the very large number of pupils that famed masters took on. Anecdotes say that some wealthy families offered munificent gifts to assure that a famous teacher would accept their offspring. They were particularly generous to masters who held high civil service posts. Others tell of eminent masters taking on poor students without pay, implying that this was exceptional. The multiplication of pupils prompted some teachers to instruct only a few senior disciples, who in turn taught the rest.[63]

Even Erudites sometimes taught privately. No doubt this practice began because more unofficial students meant more income. It became more common as the operations of the Grand Academy, like those of other organs of government, became increasingly chaotic from the first century A.D. on. Eventually, private teachers who took advanced students, although they competed with each other, no longer had to be concerned about rivalry from the Academy. From the late first century on, as the attractiveness and prospects of a civil service career diminished, private teachers came more and more to teach texts outside the official canon, including classics of divination, astrology, and medicine. In the last two centuries of the Han, teachers competed for scholastic standing in polemics, but the issues tended to be which lineage of a classic was orthodox and the proper approach to annotation, rather than substantive issues in philosophy, science, or medicine.[64]

The *ju* lineages, as we have already seen, identified themselves by association with the written classics of antiquity. On the one hand, they transmitted their canonic texts century after century with remarkable fidelity (p. 72). On the other, as each generation of scholars matured in a changing world, their interpretations of the classics evolved and ramified. The vehement attacks of Hsun-tzu on

Mencius and other teachers (to be discussed in a moment) make it clear that ju in different times and circumstances understood the tradition very differently. Direct attacks were rare; disagreements more often led lineages to assert their own positions, countering those of rivals without acknowledging them. Each side saw its aim as recapturing the authentic meaning of the canon, freeing it from the distortions of the intervening ages. By returning to the past they continually reinvented the original intent of the sages.[65]

Before the mid-third century, the Confucian pattern based on **textual transmission** was only one of several. The adherents of Mo-tzu, for instance, belonged to a severely disciplined organization, obeying its commander in their strenuous effort to promote peace by disrupting the prevalent siege warfare. This corps of activists included at least a few intellectuals. But once the fighting was over, the Mohist lineages dwindled and died out; few later scholars knew of their writings.[66] All in all, as the traditions of thought and practice of the Warring States era emerged, they were as original in the ways their adherents identified themselves with their predecessors as in their ideas. The adherents of each saw ideas as embodied in people. They thought, not abstractly of propositions, but of the teachings of a given master or a line of masters. To teach was to be responsible for not only the intellectual but the moral quality of one's propositions. For those who considered themselves Confucius's successors, as well as for others, wrong doctrines were inherently dangerous. The point was not that mistaken notions embodied unclear reasoning but that they led inevitably to pernicious action. Thus, Hsun-tzu argues that Mencius misled "the stupid, indecisive, deluded pedants (ju) of today" and, more important, offered doctrines that, if practiced, would lead to mutual destruction of the strong and the weak.[67]

This tendency of scholars to think of ideas as embodied in teachers discouraged open disputes with contemporary rivals over concepts. People generally saw an attack on an idea as an attack on its spokesman. To pitch discussion as impersonal and disinterested, in the fashion to which Greeks aspired, was practically impossible in China.

Some philosophers took the trouble to inventory their diverse predecessors and contemporaries. Hsun-tzu began this sorting out, naming a good many names, to expose what he saw as pernicious tendencies. A late adherent of the Book of Chuang-tzu tradition, for whom heterodoxy was not the issue, arrayed thinkers quite differently—and appreciatively. He wanted to show that the teaching of the true Way had been lost by his time, its traces scattered among many masters. Both of them, and others later, discussed not concepts but teachers and the doctrines associated with them.

The most curious and, in the long run, most influential inventory was that of Ssu-ma T'an (see p. 26). It needs to be read critically. Breaking with his predecessors, Ssu-ma shifted the ground, without saying so, from people to convictions. He analyzed six intellectual tendencies that were at the same time approaches to the Way and to political praxis. He named four of his six tendencies for abstractions (yin-yang, law, names, the Way) and one for the ju. He did not mention a single person aside from Mo-tzu, after whom he named the sixth. Equally odd, the word he adapted to label these six tendencies was "chia," the everyday word for "family," which no one earlier had used for thinkers or thought.[68] It unmistakably implied kinship. This borrowing had a significant corollary, for Chinese found it hard to imagine a sin worse than arguing with one's parents.

Just what chia meant to Ssu-ma he does not make clear; in any case, the topic of his survey was not philosophical disputes. It would be misleading to translate the basic sense of chia—namely, people who claim to be descended from a common ancestor—by anything but "lineage." Even so, his usage was idiosyncratic. He did not mean what chia leads anyone to expect in such a context, namely the personal relations of masters and disciples; he meant doctrines. These he discusses without referring to their transmission.

He introduced this peculiar usage to press an argument analogous to but quite unlike that of the *Book of Chuang-tzu*'s author. Five of the six tendencies have serious faults along with their virtues, but the sixth, a "lineage" of the Way (*tao-chia*), incorporates all the good points of the others. That, he argued, makes it a better guide to political practice than the imperially sponsored orthodox lineages. He abstracted his six tendencies from writings that he does not identify. Three of them, the lineages of names, laws, and yin-yang, do not correspond to any social group or collective tradition recognized before him or in his own time. They are groupings of ideas that he invented.[69] His lineage of the Way does not tally closely with the ideas of any known book; some remarks fit the *Springs and Autumns of Master Lü*, and others, the *Book of Lao-tzu*.

Thus Ssu-ma, while shifting the focus of philosophical discussion from exemplars to ideas, muddied his argument by using a term that strongly implied social relationships. Doing so encouraged muddy thinking by his successors and even by modern historians.

He was too intelligent for this to be mere sloppy thinking. It may have been his way of shifting attention away from archaic founders toward his own time—because the three traditions he invented lacked founders. The shift from proponents to doctrines may have been a means rather than an end. This word "*chia*" caught on, but his eccentric use of it did not. Naturally enough, in view of the word's literal meaning, writers after his time made it the standard term for the people who embodied or professed a teaching and the chains of adherents who passed it from generation to generation. Someone who in expounding a classic turned a tradition in a new and significant direction was said to "form a lineage of his own" (*tzu ch'eng i chia*). This did not generally mean that such a person had rejected an established doctrine to form a new one; rather, it meant that he had grown a new branch out of a tradition, like the son of a family setting up a new household.

Lineages, as the word came into common use, were not schools, either in the sense of academic institutions housed in buildings that survived one teacher's career or in the sense of "a group of like-minded philosophers," analogous to the Greek sects. If we look for lineages of textual transmission that lasted throughout the Han era, we can be sure of only one.

The ju who created the Confucian tradition organized it around texts that embodied the perfection of a past age and that thus offered a pattern for reforming the disordered present. Masters of other traditions in the late Warring States and early Han eras also evolved multiple lineages of interpretation for the same text. Han-fei-tzu, writing shortly before the end of the Warring States era, asserted that among those in his time "conspicuous for learning" were the ju (by which he meant the Confucian lineages) and the successors of Mo-tzu. He mentions eight lineages of Confucians and three of Mohists. Four hundred years later, at the turn of the first century A.D., the Han imperial collections contained, for eight Confucian classics, roughly fifty versions distinct for their texts and interpretations, each named after the founder of a lineage. Only one lineage of the Mo-tzu survived until then, and it did not last beyond the Han. Of all the other philosophical authors before the Han, only the Lao-tzu, which apparently came together late in the third century B.C., survived through the first century A.D. with multiple traditions, four of them. None of the four, so far as we know, was still active at the end of the Han. In sum, even if "school" means nothing more than a long-enduring social institution, it does not apply before A.D. 200 to any collectivity except the Confucian lineages. A belief in common descent held lineages together, not an overarching organization.[70]

In the later history of higher education in the Han, government sponsorship and the model of transmission in lineages generated **scholasticism**. The state's attempt to control curricula intensified in the first century A.D., for a growing crowd of Grand Academy graduates, each a qualified classicist, expected livelihoods. At least equally important, elite clans had begun detaching their allegiance from the weakening government and disappearing into their growing private estates. Scholarly leadership shifted away from the center. The urgency became greater as the Grand Academy became a backwater and private teachers gained in importance. Study with a well-connected master became the best hope for young men who, as time passed, had less and less opportunity for good careers in the central government. Civil service remained the ideal, but those who lacked a powerful sponsor or could not afford the increasingly onerous bribes for appointments were locked out. Scholarship and teaching came to the fore as an alternative livelihood.

The state's assertion of authority over lineages came to shape private classical studies. As officials took it upon themselves to determine exactly what authentic lineages existed and to register them, masters were eager that their own teachings be listed. The government's recognition of what it called chia-fa ("lineage models") depended on the master's civil service posts or court connections—in many instances, more so than on his intellectual influence or esteem by his peers. The most eminent masters accumulated substantial incomes from thousands and, so the record asserts, even myriads of enrolled pupils. Registration of the lineage became the key to disciples' credentials, opportunities for official appointment, and incomes as teachers. Competing lineages of the same classic proliferated, and the government did nothing to stop them.

From the first century A.D. on, ju perfected a mode of scholarship that involved writing detailed comments on individual words and phrases of classical texts ("chapter and verse"). This practice generated useful resources for teaching—some Han commentaries are still in use—but under the lineage models it became the main thing classicists did. Although it had nothing like the emphasis on logic and theology imposed by medieval European schoolmen, its obsession with commentary makes the word "scholasticism" loosely applicable.

The shift away from the primary concern with moral and political content generated respectable work for immense numbers of students who had no talent for philosophy and, as the Eastern Han period wore on, no prospects of government employment. They overdeveloped this genre to a point that some scholars found absurd: "It would take as much as twenty or thirty thousand words to explicate [a phrase of] five characters—and eventually they would be refuted. Someone would memorize a classic as a child, but not until his hair had turned white could he discuss it." There remained ample room in annotations for the didactic and even the cosmological when scholiasts wanted them, but the shift in emphasis to philology was a decided one.[71]

Governmental supervision of commentary writing was nominal, but the bureaucrats coopted the notion of education as the transmission of texts down a lineage. Lineage models became scholastic factories generating glosses endless in length and number. This institution further shaped, and was shaped by, the growing dysfunction of government and the withdrawal of elite support from it that in the third century ended the Han order. This curious educational system—which had no counterpart in Greece—evolved under central control and then collapsed with its debacle.

Deviation never became a problem of any magnitude. It was subject to an interesting tension. The personal relationship of teacher and pupil, if often distant and always formal, remained centrally important. Still, some disciples' personal principles or intellectual drives moved them in directions of which their masters unabashedly disapproved. Examples turn up in diverse coteries. The first obvious example is Confucius's own disciple Tsai Wo (or Tsai Yü), who ridiculed the master's stress on "human-heartedness" (jen) and apparently argued against the three years' mourning that was a cornerstone of Confucius's ritual teachings. His teacher's estimate of him was dismissive, but three generations later, Tsai was listed among disciples notable for their "virtuous conduct."[72] Dissenters did not make such challenges public to attract disciples, as was often the case in Greece (see pp. 124–25). In China no potential pupil would lightly decide against a connection to a lineage's senior member. But rebels did not inevitably lose their opportunities for a post. It was all a matter of whether the defection remained quiet—and teachers did not tend to advertise their problems. To study all such cases would be worthwhile, but most deviations probably remained, at worst, an embarrassing private matter.

Edicts regularly railed against organized political opposition, however friendly. The government did not have to justify harshly punishing those who challenged its policies. Surprisingly, ideologists who accused rivals of heterodoxy did not generally make a fuss about deviance from the teachings of one's master.

Kinship formed the pattern of the master-pupil relationship in **science** as well as in classical education. Here, too, in addition to practical skills, the symbolic patrimony to be maintained and passed down was one or more books. As the sciences emerged, each created a technical classic or several of them. As in philosophy, it was the obligation of each lineage to transmit intact this charismatic text that in their view embodied the wisdom, and usually the very words, of ancient sages. Each generation forged a link in that transmission. A remarkable companion of this conviction was the notion, often reiterated to the present day, that many technical innovations were boons granted by legendary rulers in antiquity—the age of perfect social order—rather than the gradual discoveries of ordinary people.

Several technical classics took the form of dialogues, often between a disciple who was a royal culture bearer (the Yellow Emperor, the Divine Husbandman) and a teacher who was also his minister of state. They dramatized the old ideal of the sage as advisor to the ruler. Such dialogues were punctilious about correct conversational form ("may I ask?"), expressing dramatically the initial puzzlement ("how can it be that . . . ?"), the master's encouragement ("what a fine question!"), and the interlocutor's relief when the answer becomes clear ("how excellent!"). These encounters took the form, in other words, of an initiation. The disciple having been permitted to memorize the classic, the teacher is explaining what his pupil does not understand (explication came after memorization, not before). The *Inner Canon of the Yellow Emperor* is explicit about the ritual circumstances. In one dialogue the transmission of the text from master to disciple and its explication requires that "an oath be sealed by cutting our arms and smearing blood" after three days of purifications.[73]

Were there distinct lineages of medicine? Yes, beginning at least as early as the doctrinal writings of the *Inner Canon*, which probably accumulated over much of the first century B.C. By then, physicians were using cosmology to structure their doctrines (p. 270). By the end of that century, each of three separate lineages—named for the Yellow Emperor, the legendary physician Pien Ch'ueh, and a Master Pai—had produced collections of short texts. The final content of the extant *Inner Canon* came from at least two of these three. At some point, in other words, the borders between lineages broke down. In the Han some people received texts from—that is, were initiated into— more than one medical lineage. The biography of Ch'un-yü I in *Records of the Grand Scribe* makes it clear that he was one of these disci- ples of more than one teacher, and that one of them transmitted to him the writings of more than one lineage.[74]

The scholars who studied the other sciences also organized them- selves around written revelations. There, too, the survival of a single classic reflected the dominance of one line. The point of departure for materia medica was the *Divine Husbandman's Materia Medica*. For astro- nomical mathematics we have the *Gnomon of the Chou*, which begins with a dialogue between the duke of Chou and one of his advisors. The oldest extant alchemical work, the *Nine-Cauldron Divine Elixir Canon of the Yellow Emperor*, is perplexing. It begins with the Yellow Emperor's initiation by a divine lady. The remainder is in the form of a straight- forward treatise, but, as the title implies, the whole is supposed to have been revealed to the reader in turn by the Yellow Emperor. The body of the treatise does not mention the Yellow Emperor, however, or any other ruler. The revelatory framework, including the title, was probably imposed after the Han on a Han original that mentioned no legendary figure at all.[75]

Mathematical Methods in Nine Chapters, like the classic of materia medica, used straightforward exposition instead of the archaistic dialogue form. Nevertheless, both came before readers with prefaces that stressed their links to the sagely beginnings of culture. Here is that of the mathematics book: "In antiquity the emperor Fu-hsi first drew the eight trigrams of the *Book of Changes* in order to be in touch with the power of the gods and to set out in their proper categories the actualities of the myriad phenomena. . . . The duke of Chou, in codifying the rites, set out nine basic types of mathematical problem. The outcome of those nine problem types are these nine chapters!"[76] This reasoning may seem far-fetched, but it reflects the desire of some commentators and some readers to keep the connection with sagely revelation intact.

That was the picture as the classics presented it and as Han readers saw it. Critical dating puts all these writings in the first century B.C. or later.[77] Their compilers or later authors of their prefaces attributed them to the sages of high antiquity in order to provide respectable origins for new technical traditions. This was not forgery or any other defect, as it might seem from today's point of view. A legitimate tradition of inquiry, however new, needed to belong to a lineage to be taken seriously. The line of transmission had to be linked to the golden age, if not literally descended from a sage who lived then. Attributions to the Yellow Emperor or the Divine Husbandman were a proven way to assert significance.

The Chinese answers to the questions raised about collectivities and membership are now clear. Membership in a lineage of masters and disciples was fundamental to higher education, even though schools in any normal sense of the word were inconsequential. Pupils took on a commitment to transmit a written canon. The teacher-student relationship required deference to the teacher, but not sole allegiance. The central government increasingly influenced these lineages as the Han progressed, and their focus shifted from statecraft and self-cultivation to philology.[78] Because the lineage-model system did not codify the conduct of students or masters, the highly ritualized character of teacher-pupil relations remained the main instrument of control. Deviation, so far as the record reflects it, was rare. It was not an important issue, and cases were resolved individually and privately.

Text-oriented lineages in science and medicine emerged from the first century B.C. on, adapting the forms of philosophical classics to traditions of technical practice. The extent to which the state shaped them depended on its involvement in the particular field.

Arguments, Books, Commentaries, Memorials

Chinese philosophers and scientists left many types of writing, some of which Greek philosophers and scientists also created but used in quite different ways (p. 118). The most important Chinese types were classics and canonical collections of them, memorials to the throne, dialogues, treatises, and commentaries. The Greeks did not write classics, canons, or memorials.

Certain forms of **argument** were characteristic among Chinese. We will pay attention to which were prevalent, what they disagreed about, and how the general desire for an orthodoxy affected topics of disagreement. Examining, further, who adjudicated, how that affected the form of argument, and what tipped the balance between disagreement and consensus, and then surveying books and memorials (a bureaucratic form), will cast light on the circumstances and ideas of cosmology and the sciences, and ultimately on reasons for differences in the two cultures.

It has often been asserted that when science evolves, it, like philosophy, evolves through argument. But there are many kinds of argument. The dictionary insists that "dispute" and "debate" are special types of argument. "Dispute" basically refers to "a contradiction of an assertion and implies vehemence or anger in debate"; "debate" implies "a formal argument, usually on public questions, in contests between opposing groups," and hence usually refers to face-to-face oral competitions, even between individuals. "Debate" is also generally used for "polemics"—reiterated, usually written, exchanges between opponents. The most germane Chinese word, pien, can in different contexts correspond to any of these senses. It can refer to dialectical activity of various kinds, from written disputes with predecessors long gone to informal oral arguments with fellow disciples of the sort seen in the Confucian *Analects*.[79]

In the six hundred years that interest us we find a good deal of argument: overt disputes—much less common in philosophy and science than in political decision making and discussions of orthodoxy—and debates on technical topics, which were conspicuously rare. Disagreement, fruitful though it was in testing arguments, was mostly muted or indirect. There is no record of public philosophical arguments in ancient China; most of the few debates took place in palaces—precincts strictly forbidden to anyone not ordered to be present. The *philosophic* focus remained on writing; for instance, when Wang Ch'ung, the critical philosopher of the first century A.D., ranked his predecessors, he did not mention any oral encounter.[80] One might object that books lasted and memories of debates did not; but such memories were recorded in Greece, and accounts of oral disputes on policy were generously preserved in China.

Not many Chinese intellectuals were attracted to open conflict of any kind. There were several reasons. Most thinkers were cultivated offspring of aristocratic households. That many refused official appointments in the catastrophic second century A.D. suggests unearned income—which is not surprising, for by that time much wealth had gravitated from the poor to the rich. Some earlier intellectuals did not have to earn a living. But gentlemen who aspired to careers made their way in systems first of patronage, then of civil service, and finally of registered private lineages. None of these forms encouraged confrontation.

The few recorded oral arguments not concerned with current affairs, orthodoxy, or ritual came early in our period. Aspirants to patronage during the late Warring States era occasionally instigated them in aristocratic courts, hoping to make a splash. The connection of display with livelihood accounts for the seeming resemblance of verbal attacks in the age of patronage to the debates by which Greek philosophers advertised themselves to potential pupils. But the Chinese form of this intellectual aggression tended to be quite different.

After unification, what most gentlemen hoped for was not celebrity but secure status and political influence. Philosophers without official status still tended to address their arguments to rulers, and scientists to take up matters of governmental concern, but neither saw any need to develop running debates. Authors of both kinds often expressed their views forcefully, even when their views conflicted with those of predecessors, but they did not often do so in a confrontational way.

Let us look at four illuminating comments on argument before 200 B.C., first philosophical and then political, embodied in real or supposed circumstances:

1. One of Mencius's disciples asks him—not in public—an embarrassing question: "Outsiders call you fond of argument (*pien*). May I ask what this is about?" Mencius replies, "How could I be fond of argument? I have no choice." Briskly surveying the history of the world, he concludes that he lives in an age of social, political, and moral decay, when "gentlemen with no official post express their views without inhibition" and echo the worst philosophical models (such as the Mohists). Wishing "to correct people's minds, stop

heterodox doctrines, oppose extreme actions, and banish licentious words—to carry on the work" of Confucius and other sages—"I have no choice."[81]

The nub is that dispute is undesirable and is necessary only in bad times like the present. In a degraded age, others' willingness to state their minds without regard to hierarchy forces him to take the offensive. For Mencius, however, this is not a mere matter of winning an argument. He is not just one more debater, because he is carrying on what he sees as the work of Confucius's authentic successors: suppression of unorthodox actions, words, and thoughts. That Confucius did not share this obsession does not deter him.

2. The group of anecdotes most frequently cited as historical instances of philosophical debate is associated with Kung-sun Lung, famous in the late Warring States era for his paradoxes. In the best-known story, the philosopher of history Tsou Yen, while visiting the court of the lord of P'ing-yuan, a celebrated patron, put down Kung-sun, a client there. The latter was renowned for his argument that "a white horse is not a horse." We might see this today as an attempt to explore the difference between the set of all horses and the set of white horses, but at the time the novelty of the argument aroused much indignation. When the patron asked Tsou to comment, he asserted flatly that such propositions are not among those that should be permitted "in the realm"—that is, by the authorities. Rhetorical tricks "do harm to the great Way. This tangled-up wrangling to talk the longest can only bring harm to gentlemen." At the end of the story "all those sitting there praised [Tsou's speech]." This is one master's critique, not of another's doctrines, but of the dialectic enterprise. The universal praise implies that Kung-sun Lung and his disciples were absent. Tsou won his point by resorting to exactly the oratorical flash he claimed to condemn, adding a hint about heterodoxy to clinch the issue.

Only one account in this cluster is about a face-to-face argument. "K'ung Ch'uan and Kung-sun Lung argued in the habitation of the lord of P'ing-yuan. In the course of their intense disputation, they came to 'Jack has three ears.' Kung-sun argued this proposition with many fine distinctions. K'ung did not reply; after a few minutes he made his excuses and left." Later in the narrative K'ung revealed to his patron that he had aborted this debate over a stock verbal paradox because he was unwilling to take seriously an untrue assertion. This second story, though a rare account of a philosophical debate, resembles the first in its disapproval of free intellectual exploration.[82]

3. A chapter in the *Springs and Autumns of Master Lü* on respect for teachers gives oral argument a certain place in education: "From time to time argue about interpretation in order to evaluate the underlying meaning. Do not dispute frivolously; it is essential to be centered on the correct method. If you succeed, there is no reason to be proud; if you fail, there is nothing to regret. In either case, you are certain [eventually] to return to the basis [i.e., succeed in practice of the Way]."

Argument, again, was not a pervasive aspect of scholarly life. It was permissible as an occasional tool to throw light on the text, but it was not permissible to wander wherever the intellect leads. Whether this exercise, which could be dangerous if high spirits take over, was precisely a dispute is questionable. It would certainly not be preparation for a Greek public free-for-all.[83] A few masters encouraged argument among pupils as a teaching tool, but disciples did not lightly entertain the notion that their teacher was wrong, and practically never said so.

4. Written one-way arguments have their place in philosophical books. Some disputants are overt, and some do not acknowledge with whom they are disagreeing. Still, Chinese arguers take exception to people—mostly dead ones—not to disembodied ideas. A particularly obvious example is the condemnation of the ju in the fourth-century B.C. *Book of Mo-tzu*. The Mohist argues in a series of anecdotes that Confucius was a hypocrite and that the contemporary scholars who claimed him as a model were even worse hypocrites.

Confucius's successor Hsun-tzu (d. ca. 238) goes even further, attacking by name a dozen philosophers. He combines in this group of undesirables non-Confucians and those he judges to be unorthodox ju among his predecessors and contemporaries. Prominent among the ju are Mencius and Confucius's own grandson Tzu-ssu, both of whom lived well before Hsun-tzu's time and may be predecessors in Hsun-tzu's own lineage. His criticisms are epigrammatic, no more meant to initiate exchanges with his living targets than with the dead. On the other hand, a large portion of another chapter purports to transcribe a confrontation in the palace between Hsun-tzu and a commander on the principles that underlie the use of military force. It begins with a single credible exchange. Then the commander abruptly becomes a straight man, repeatedly asking Hsun-tzu for his views, which the philosopher states at length. Finally the military man disappears, and the account lapses into straightforward essay form. This is not even a serious imitation of a debate. It is, like our second example, a form of entertainment for courtiers.[84]

These examples confirm that the few philosophers who indulged in open argument preferred one-sided written polemics in philosophy and reserved face-to-face confrontation for politics. Although records of debate on cosmological or ethical principles in the Han are rare, we have evidence, from our whole period, of conferences in which high officials and scholarly experts discussed matters of policy, ritual, and orthodoxy. The histories document seventeen formal imperial conferences on such matters in the Western Han and twenty-six in the Eastern Han. The most famous, because the best documented, were those of 81 B.C. and A.D. 79, the first on fiscal and foreign policies and the second on discrepancies in the interpretations of orthodox canons. Exactly how certain texts ought to be read was an important issue for debate in the second.

In both conferences the emperor delivered the verdict, and in both the practical outcome of the consultation turned out to be insignificant. Records of the two leave no room for any misconception that the judgment or the eventual result depended in any simple way on the intellectual merits of the presentations.[85]

Returning now to philosophical arguments, what do we see when we cast an eye over those for which we have records? Most reflect a tension between the two worlds in which clients and officials were immersed: their social milieux, which discouraged the open expression of disagreement, and the court, in which a decision by the ruler or his high officials on an exigent matter was likely to be more competent if they were willing to consider others' points of view.

Philosophers, though important as consultants and occasionally appointed to official positions, were not likely to debate philosophical, physical, or medical matters in the palace. When philosophical topics came up at court, the issue was likely to be definitions of orthodoxy or accusations of unorthodoxy. Jockeying for control of official appointments occasionally led to intense attacks on rivals. The issues were chosen for their political resonance, not for their conceptual interest. Thus a group of ju between 140 and 136 B.C. pleaded successfully that "depraved teachings"—"every doctrine not within the scope of the six arts, the methods of Confucius"—should be wiped out. If the Grand Academy taught prospective officials only a single set of doctrines, it would be possible "for the system of law to be unified and the statutes and institutions to become clear." Just such arguments led to exclusive support for the five classics in 136 B.C.[86] But they were in no sense part of a discussion meant fully to air opposed points of view.

This becomes even clearer if we consider what was probably the most vehement confrontation of the second century B.C. It involved two of the most important philosophical figures of the time, Liu An, king of Huai-nan, and Tung Chung-shu.

Liu, aware that officials were pressing his young nephew the Martial Emperor to get rid of the local kings once and for all, cited natural portents in his *Book of the King of Huai-nan* (139 B.C.) to argue against the drive toward a state orthodoxy, to blame the social conflict of the time on evil scholars manipulating the ruler, and to plead with the monarch to forsake them. He failed. Tung was prominent among the enemies of the imperial relatives. In a memorial of 135 B.C., generously citing classical precedents as well as omens, he warned his monarch that it was an urgent matter to establish a clear moral hierarchy based on a single set of principles. This program left no room for alternative ideals. In pursuit of his high-minded goal, Tung forcefully advised curing the ills that "arrogant, extravagant" imperial relatives caused. The therapy he prescribed was for the emperor to have large numbers of his own flesh and blood executed. In the end it was a disciple of Tung's who ordered Liu's execution. This dispute was so carefully phrased in ethical generalities that most historians of philosophy have failed to note it.[87]

What we conclude from these examples is that although many Chinese intellectuals were as able to hold their own in an argument as anyone elsewhere, the values of their culture stressed harmony and consensus. Those values did not force them always to agree, but they did motivate most intellectuals to express rivalry and contention judiciously and indirectly, except in circumstances that called for open disagreement. The circumstances at the top of Chinese society made the ruler a natural intermediary for those who had access to him. Although many occidental readers today will find this veiled prudence less congenial than the Greeks' open aggression, a preference either way is irrelevant to understanding differences in what shaped cosmology, science, and medicine.

One way Chinese learned to express disagreement was through **scholastic competitions**. In the first and second centuries A.D., orthodox teachers used a game, usually called stumping, in which students displayed competitively, but without disruptive bickering, their command of classical annotation. The term is mainly known from Discourses Weighed in the Balance (late first century B.C.), written by the teacher Wang Ch'ung: "After the Han dynasty established the office of Erudite, masters and their disciples engaged in stumping, with the aim of fully exploring the depth of the Way and delineating the principles of argument. If they had not posed unrestrained challenges, they would not have obtained concordant explanations. If they had not quizzed each other in bitter earnest, they would not have heard sweet answers."

The passage does not say clearly what stumping was, but it definitely was not argument or debate. The biography of one Tai P'ing (early first century A.D.) gives some crucial details using slightly different terminology. The setting was the court, and the object was not fathoming the Way (which is how Wang saw education) but intellectual entertainment. When Tai was a young civil servant, the Radiantly Martial Emperor invited him to a gathering of high officials and "had him engage with the scholars in challenging each other at 'explication.'" Tai excelled at this and was appointed a confidential advisor to the emperor. Later on, "at the congratulatory audience held on New Year's Day, all the high officials were in attendance. The emperor commanded those ministers who could expound the classics to 'quiz' each other. Whenever someone did not know the meaning of something, his sitting mat was taken away and added to those of the one who did. Eventually Tai was sitting on [a pile of] more than fifty."[88]

This game encouraged the display of erudition for collective amusement and perhaps spurred learning. It does not resemble any Greek usage, but it does call to mind the scholastic disputation that was one of the most fundamental exercises in the medieval European universities. The schoolmen disputed a wide range of questions, with each student marshalling arguments, both pro and con, from authoritative sources, including ancient books and Scripture, before drawing the conclusion he favored. Again the resemblance is superficial; players at stumping simply had to explicate what a word or phrase meant in a given classical text. It was, in other words, extemporaneous oral "chapter and verse" scholarship.

The **dialogue** form began in Chinese philosophy and persisted in science (as we have seen, p. 59), surviving into the age of the treatise. The dialogues of Confucius's *Analects* comprise a great range of dialogue types: those between master and disciple, between peers, between ruler and philosopher, and between philosopher and an outsider who wants help or enlightenment. The *Book of Chuang-tzu* sets down encounters of grotesquely deformed sages with chance interlocutors, new widowers with visitors to the wife's funeral, emissaries offering high appointments with hermits who spurn them, Confucius with a bandit whose code of thievery is more exacting than the sage's code of virtuous conduct. All of these, like some of Plato's dialogues, are monologues with listeners inserted for dramatic effect. The *Chuang-tzu* is less straightforwardly didactic than the *Analects*, for the point of many of its putative arguments is the delusory nature of what conventional people consider right thinking, and the uselessness of argument as a way to find out what is true. In other words, Chuang-tzu demolishes all partial views in the hope of shocking people out of their hackneyed convictions and into a mystical vision that unites opposites. Like other rational arguments meant to communicate the trans-rational and indescribable, his are deeply paradoxical.

Dialogues between legendary rulers and their ministers and other political motifs figure regularly in technical books that draw on classical literary forms. The authority of the theory of monarchy embedded in the homology of state, body, and cosmos attracted teachers of scientific traditions. Even when they were not employees of the state and even when their subject was mathematics or medicine, their dialogues insistently remind readers that correct thought radiated from the center. They were very different from the face-to-face debates of the Greek agora.

One reason that treatises on science and medicine appeared late in China is that treatises appeared late—if by treatises we mean **books** of substantial length on one topic, written at one time. Physical form decisively affected the evolution of the book, a story that is now gradually emerging from the study of excavated writings. This theme in material culture is of the first importance for understanding Chinese intellectual collectivities because they so consistently organized themselves around a written classic. The survival of such a book required careful maintenance, usually by a collectivity; reciprocally, holding a lineage together for generations depended on dedication to a stable scripture.

Most authors in the period that interests us wrote with brushes on long, narrow wooden strips, each of which usually held one line of text. They strung them with cords into sequences that could be rolled up for storage and unrolled for reading, rather like bamboo shades. Large bundles were so unwieldy that the practical limit amounted to hardly more than ten pages of modern print. When the cords that bound a bundle rotted after long storage, the strips easily became jumbled; it was often impossible to sort them back into the correct order. Authors could extend a composition over several bundles of strips, but they usually designed their writings to fit into one bundle or less. Silk was too expensive for most users, and paper did not begin to replace wood for the purpose until after the Han era.[89]

Many manuscripts unearthed since 1970 have shown us what Han books looked like.[90] Finds in the tombs of nobles have included some on large pieces of silk, rolled up or folded, but these are luxurious copies of works normally composed and copied on wooden strips.

What we are used to calling a book began as a number of distinct writings attributed to the same author or on the same topic that someone had accumulated. Two teachers, former pupils, or aristocratic collectors were likely to own different bundles. Diverse short texts, texts copied together, even multiple, slightly different manuscripts of what later became the same classic, have turned out to be the norm in excavated tombs. If no one integrated the writings into books of fixed content, such works were likely to drop out of circulation.[91] The contents of a number of recently disinterred medical books and divination manuals, for instance, were previously unknown.

The ju, who revered books as links to antiquity, gradually compiled an authoritative sequence of bundles and passed it down. Those who claimed membership in the lineage of Confucius were preoccupied with the classics that they believed had passed through the hands of the master. The ju lineages could determine the content of books because they were organized. Their ability to organize added a great deal to the authority of their writings. For instance, we can now see that between two and eight generations of disciples wrote the constituent parts of Confucius's *Analects*, little, if any, of it from verbatim transcripts. Ju began compiling them from perhaps the mid-third century B.C. on (different lineages put together different sequences).[92] Once editors fixed the text of a classic—long after its earliest parts were written—ju scholars were able by great discipline in copying and cross-checking to maintain it with remarkably little alteration over century after century before printing made that easier.

Other traditions of scholarship, in particular scientific and medical ones, began only after ju had established this pattern of compilation and transmission. Lacking continuity of organization, they had to cope with unstable canons. In the *Inner Canon of the Yellow Emperor*, assembled from short texts in the first century B.C. or a little later, the editing was light enough to maintain many traces of original diversity. Some parts contradict other parts, quote them, explain points in them, or explicitly argue against them (see p. 199). Two distinct but similar versions of the book, fixed in the seventh and eighth centuries, neither markedly more pristine than the other, have survived to the present.[93]

At the end of the first century B.C., the process of editing these jumbled accumulations to make books with fixed content gathered momentum.[94] Two bibliophiles, librarians, and philosophers,

Liu Hsiang and his son Hsin, assembled great numbers of books from odd bundles they foraged. They were responsible for the ultimate forms of much of the surviving early literature, including records of astronomical observations and other writings of scientific interest.

In thinking about written classics and treatises in our six centuries, then, it is best to remain aware that books came into existence, by fits and starts, much later than in the Greek world. Before the first century A.D., the notions of a book as the product of one author and of composition at a specifiable time do not yet generally apply. It makes better sense to view most of the first scientific and medical treatises as assembled over a generation or more by members of a lineage, who sometimes drew on texts that originated elsewhere. The content of the final book depended as much on which short texts happened to be available as on rational selection.

The formation of **canons** in the broad sense—groups of classics recognized by some authority—began with the Han era and continued through history. This step was fundamental to the move from the period of the classics to that of classicism. Like the formation of canons in Jewish, Christian, and Islamic cultures, it was an outcome of social crisis and intellectual dislocation.[95] The Chinese five classics were a typical model for orthodoxy. This collection guided personal conduct and state policy, contained information about every aspect of experience, and provided a model for writing. There was no canon of this kind in pre-Christian Greece.

Canons, like individual classics, were an invention of the ju lineages. Most of the books that they made classics began as common property. After 136 B.C., when various ju persuaded the emperor to give unique status to the five they favored, other classicists eventually ceded proprietorship to them, as we have seen.

Canonical books differ from others in more than the authority that designates them. Their sponsors claim for them the inner harmony and coherence of the golden age. Understanding them right meant seeing past a disordered surface. The *Book of Changes*, for instance, looks to the naive reader like a grab bag of divination texts ("Divination inauspicious"), bits of proverbs ("Moon past full, horse sure to be lost"), and snatches of songs ("A crane singing in the shade . . ."). One of the great projects of orthodox scholarship from the Han on was reinterpreting the *Changes* as a handbook for morally guided action. The lines about the crane, for example, came to be "really" about the involuntary influence of a cultivated gentleman's inner being on kindred spirits.[96]

Long, more or less systematic **treatises** on a particular subject appeared later than collected teachings of individuals or compilations in dialogue and other forms. The earliest reasonably coherent book in expository form that is authentic, was plausibly written at one time, and is not an imitation of or interpretation of a classic is the *Mohist Canons* of the late fourth century B.C., which contains much of interest for physical science.[97] Just as striking, by about 100 B.C. there was only one other full-length book that meets the same criteria, the complex and tightly integrated historical treatise *Records of the Grand Scribe*, an important source for astronomy, astrology, and medicine.

The form and authorship of scientific books evolved in step with the general evolution of books. The catalogue of the imperial library, compiled in 6 B.C., included nearly six hundred titles. Of these, forty were on astronomy (considerably fewer than the seventy or so on divination), eighteen on medicine, the same number on sexual and other techniques for lengthening life or becoming immortal, one that may be on siting, and none on alchemy. Almost all of these books have disappeared.[98] A handful of books of the time, mostly not in the catalogue, have been passed down over the centuries. These and another few among recently excavated manuscripts make up the odd assortment of technical literature that has survived.

Of the earliest extant medical books, excavated from a tomb of 168 B.C. at Ma-wang-tui, Hunan province, the *Moxibustion Canons* are short individual writings (or versions of the same writing) rather than a compilation, and *Formulas for Fifty-Two Ailments* is a diverse collection of drug and ritual therapies, perhaps the personal accumulation of one healer. These do not contain explicit rationales or doctrines. The *Book of the Pulsating Vessels*, buried about the same time in Hupei province, writes of pathological ch'i as agents of three yin and three yang disorders.[99]

The earliest technical books that build up elaborate doctrines, notably the *Inner Canon* and the *Gnomon of the Chou*, came together in the first century B.C. or a little later. The former mostly, the latter partly, are in the dialogue form of classics. The *Mathematical Methods in Nine Chapters* of around A.D. 100, in contrast, is a compendium of problems with solutions. In the next century, as the stylistic influence of the classics weakened, more technical handbooks appeared, such as the *Divine Husbandman's Materia Medica*. But the *Canon of Eighty-One Problems in the Inner Canon of the Yellow Emperor* takes the form of an initiation dialogue, echoing more systematically the form of the *Inner Canon*, which it explicates.

Commentaries, as in Europe, India, and eventually the Islamic world, became a characteristic form of scholarly activity once philosophy turned to look back at the classics. In the Western Han era, when most thinkers were obsessed with their predecessors, imitations and interpretations of orthodox texts became important genres. Full-dress commentaries, if by that we mean running glosses on words and notes on their implications, came into their own from the first century B.C. to the second century A.D. as scholiasts perfected and wielded the chapter-and-verse genre. This separation of scholarship from both philosophy and statesmanship in the Eastern Han era was related to state control through registration of lineages, as we have seen. Like some textually fixated modern academic specialties, Eastern Han scholasticism offered its adherents highly respectable employment, access to pupils, tidy reputations, and, occasionally, modest fame for some prodigy of painstaking but not necessarily useful explication.

Once commentaries became the dominant genre of orthodox scholarship, they served for the rest of the Han era as the normal medium for expressing anything original. As in earlier forms, innovation and awe for antiquity sat side by side. Because annotators generously quoted the literature available to them, they preserved parts of books that were soon lost. The *Divine Husbandman's Materia Medica* was one of these. Textual scholars from the twelfth century on have assembled thirty-eight reconstructions of it from fragmentary citations and commentaries in later books.[100]

Not all classics attracted critical study in the Han era. The *Mo-tzu*, once its lineages of transmission died out, lay almost unread until rediscovered by scholars at the end of the nineteenth century. Yang Chu (fourth century B.C.?) may or may not have argued for individualism; the quotations of his writings in early sources are too scanty to allow a firm conclusion.[101]

A curious tension characterizes the assumptions that underlie commentary, but it is not a competitive one. Texts attract exegesis when scholars become convinced that change has obscured their meaning or that they are in some sense damaged and need repair. They assume that the text was originally a perfect, entirely comprehensible whole. But consider the *Inner Canon of the Yellow Emperor*, which came together as a rather incoherent accumulation of short texts. When the *Canon of Problems* ingeniously explained away its most important inconsistencies, the neatness of the outcome contradicted the disorder of the *Inner Canon*. The later book does not admit that the earlier one contains a great many contradictions; in its dialogues a master simply provides clarifications one by one as a disciple requests them. Subsequent treatises and commentaries had to cope with the differences between the untidy *Inner Canon*, the tidy *Canon of Problems*, and the never unanimous tidying-up of later systematizers. Modern scientists avoid this problem by assuming that new knowledge trashes the old, leaving only one state of the art. But in ancient China all the knowledge of antiquity remained co-present, the older the more inherently valuable.

Memorials as a genre had no Greek counterpart. They were official, formal communications to the emperor or to the central government in his name. Addressing the ground beneath the emperor's feet rather than his august person was not a mode nicely adapted to philosophizing. Still, memorialists often argued vehemently against opposing points of view on policy and related matters.

Memorials regularly played crucial parts in decisions about important intellectual matters. They were in principle petitions or submissions rather than contributions to discussion, but they could be part of complex interactions. Some were responses to inquiries about policy from the emperor. The best known are the three by Tung Chung-shu, probably written in 130 B.C., arguing that the emperor should restrict all official appointments to initiates into "the arts of Confucius" and end all other proprietary Ways. Tung's proposal was not put into practice. The ruler took what he wanted from memorials without needing to respond or to justify his choice.[102]

Experts also submitted technical reports for the emperor in the form of memorials. Some are documents of prime importance for the history of science and medicine and are quoted in the dynastic histories. An edict of around 176 B.C. ordered physicians skilled in prognosis to summarize their strengths, their teachers, what books they had been taught, which books they had transmitted to others, where their most prominent patients lived, and so on. The detailed reply of Ch'un-yü I (see p. 59) gives us the earliest account of the qualifications of a doctor who was not a medical official. In his circumstantial account of his experiences in various courts he mentions a couple of disagreements with rival physicians (one of them his patient). None led to an argument between the principals in the presence of others. Another example, one of the most revealing sources on Han astronomical thought, is a sequence of discussions beginning around A.D. 68 in which officials analyze shortcomings in the official system for computing the calendar and advise on policy for dealing with such problems in the long term.

In these instances, as in many already mentioned, the final decisions came from the throne. In four cases, the ruler ordered a general conference of high officials to deliberate on the issues. A remarkable account of A.D. 175 deals with a charge by two officials that errors in the current astronomical system were responsible for a rebellion (in Chinese astrological doctrine, rebellions, like eclipses or earthquakes, could be portents). An edict ordered high palace officials to forgather in the Office of the Minister of Education for one of the imperial conferences mentioned above (p. 66). A participant's record describes the formal array in the courtyard, "the Commandants facing east, Palace Attendants and Leaders of the Court Gentlemen, Grand Masters, Officials with salary of a thousand bushels and of six hundred bushels of grain in their ranks facing north, Court Gentlemen for Consultation and Erudites facing west, with a Clerk of the Civil Affairs Section seated in their midst to read the imperial edict." A particularly eminent Court Gentleman for Consultation (remarkably) debated the two complainants, one an insignificant court official and the other a provincial Accounts Assistant. They were definitively outranked and outnumbered. Because their charge involved a portent of misrule, casting doubt on the Mandate of Heaven (the emperor's divine right to govern), three high officials accused them of disrespect for the emperor and recommended that they be sentenced to manual labor. The topic of discussion decreed for this conference was not simply their accusation but calendrical administration. Cowing the lowly astronomical critics demonstrated the gravity of the meeting and set the policy discussion in motion.[103]

Except for memorials, the kinds of communication that were important in the evolution of science and medicine in China were largely analogous to those in Greece. But the analogues are superficial. In China there was no tradition of public debate of the kind that was central in the Greek world. Philosophical and scientific argument tended to be written and indirect and was seldom confrontational. The scholastic competitions that had a short vogue in Eastern Han China were not arguments over ideas but a matter of orally footnoting classical texts in private company.

Even the dialogues, treatises, and commentaries, which have Greek counterparts, are strikingly different. Their content reflected and influenced the social settings of scholarship as they evolved from patronage to official employment to the lineage-model coteries of the late Han. Ideas related to the state orthodoxy evolved similarly as members of the literate elite formed and responded to it. Education rooted this set of ideals within individuals, at the same time encouraging an important role for macrocosm-microcosm themes and other political leitmotifs in scientific and medical as in other writing, a matter at which we will look closely in Chapter 5. Varying degrees of governmental involvement in each technical field affected, but did not determine, how far state ideology shaped its content.

The Social Nexus of the Sciences

People at many levels of society did technical work in ancient China, but almost all of the data that bear on the cumulative traditions of science and medicine tell us only about members of the tiny elite descended from aristocratic clans. They were not much like those who, for instance, cared for the health of the overwhelming majority of farmers. Whether these healers were herbalists or popular priests, they tended to pass down their technical skills from father to son or from mother to daughter. This not as likely to be the case for gentlemen, even those, like Ch'un-yü I, who inherited official sinecures (Ch'un-yü's official biography calls him by his title Director of Granaries).

The few gentlemen who became cosmologists supported themselves at first through patronage: local potentates collected clients whose skills might help their states survive. The rulers did not, so far as we know, value systematic rational thought for its own sake, but found it, like the stocks-in-trade of diviners, diplomats, and tricksters, a potentially useful resource. With the unification of China, official appointment gradually provided a more stable form of livelihood, although intellectuals were only a small minority of those it supported. People who made careers as astronomers, mathematicians, or physicians, as in the earlier patronage system, became inured to a pattern of communication in which they offered advice and proposals with no certainty that their superiors would consider or or even acknowledge their suggestions. When classical annotation gained prestige from the first century B.C. on, the government began to register lineage models of masters and disciples. This official gesture, though more often a matter of graft than of actual supervision, encouraged the mass movement of private scholars into scholasticism of a highly stereotyped kind. This movement was one facet of their retreat from political involvement as central authority decayed. The scientific and medical literature that emerged at the same time no doubt attracted some scholars as preferable to a career in commentary writing.

These changing career patterns over six hundred years affected members of an elite that valued harmony and tended to worry that disagreement would verge on heterodoxy. Although disagreement in fact played fruitful roles in every department of thought, it did so indirectly and in most cases without overt confrontation between contemporaries. The personal attacks that occur in the writing of a few Warring States thinkers are best understood as instances of the prevalent tendency to think of ideas not as abstractions but as embodied in masters. The Chinese mirror image of Greek public debate was a tendency to seek agreement and to claim it even when it did not exist. Even when interpretations conflicted in the chapter-and-verse commentaries of the Eastern Han, the tone remained genteel and impersonal. Acerbity entered when the issue was orthodoxy, seldom otherwise.

3 The Social and Institutional Framework of Greek Science

The Origins of Investigators, Employment, Patronage

The problems we face in analyzing the social and institutional framework of Greek science are the same as those we identified at the start of Chapter 2. Our chief questions are these: What strata of society did philosophers and scientists come from? How did they earn a living? Did their work as philosophers or scientists help them to do so? How far did they depend on patronage? Did that affect the way they defined and pursued their inquiries in different fields and different periods? To interpret the concrete evidence for particular Greek philosophers and scientists or groups of them requires setting out some of the general features of ancient Greek social structures. We may take as our starting point some of the categories the Greeks themselves recognized as important.

Throughout Greco-Roman antiquity the first and fundamental social division was between *slave and free*. Not all slaves were born into servitude. Some became slaves, temporarily or permanently, through defeat in war or personal misfortune. Diogenes Laertius reports that Plato at one point was sold into slavery on the orders of Dionysius I, although he was ransomed immediately.[1] In the fourth century B.C. the question of whether the institution of slavery is "natural" was disputed—for some held that slavery is always and everywhere a matter merely of law or convention (*nomos*). Aristotle refers to that view but rejects it.[2] For him, as for most Greeks, slavery was as natural as ruling. But even those who thought slavery a matter of social, not natural, distinctions made no move to abolish the institution.[3]

The free included more and less aristocratic and more and less well-to-do families. One initially very important difference was that between the noble (*eugeneis*, "wellborn") and the rest. The *eugeneis* did not owe their status to grants of fiefs and titles by rulers as Chinese nobles did; rather, they were families who prided themselves on their distinguished ancestry. In the archaic period, of the seventh and sixth centuries B.C., much of the power and wealth in many Greek states was in their hands, but with the rise of the institutions of the city-state at the beginning of the classical period (from the end of the sixth century B.C.) their influence declined. This came about first because of changes in the styles of warfare. The use of hoplites (heavily armed infantrymen) in massed formation meant that victory in battle depended less on individual prowess than on their disci-plined coordination. Their increasing military importance weakened the grip the ancient families had on political power (as analogously it did in China). Second, tyranny undermined the position of the aristocrats and, paradoxically, favored an eventual more even distri-bution of wealth. Third in Athens in particular, Cleisthenes broke the power of the ancient families in the Athenian tribes when, in 508 B.C., he reorganized these on a purely geographical basis.

In the fifth century B.C. and later there were still hereditary "king-ships" and "priesthoods" in many city-states, Athens included, but—with the notable exception of Sparta—the former were of little political significance and tended, like the latter, to be merely hon-orific and ritual positions. In Athens the key political office was that of general, and the ten of them were elected yearly by the citizen body as a whole in the Assembly. However, some of the ancient families still retained considerable wealth, standing, and influence. One such was the Alcmeonid clan, to which two of the most influential Athenian leaders—namely, Cleisthenes in the sixth and Pericles in the fifth century B.C.—belonged (the latter on his mother's side). Evidently, even for potential leaders of the democracy it was no disadvantage to belong to an aristocratic family; quite the contrary, thanks to the connections it brought.[4]

Birth, then, provided one basis for differentiation within the free, and wealth was another (more on that below). The extent to which occupation was a third is less clear. It is true that Greek political theorists often suggested divisions within the state based on occupation. Plato has his three clearly segregated classes in the Republic—the guardians, the auxiliaries, and the workers—and within the last he insists that each craftsman should undertake just one activity. Aristotle, too, analyzes the main parts of the state in terms of farmers, craftsmen, traders, manual laborers, and so on, classifying different constitutions according to the roles and rights each group had in each. In his idea of the best constitution, citizens are not either farmers or artisans or traders. He thought those occupations were incompatible with the best life, for that included full participation in the time-consuming processes of political decision making, attendance at the Assembly, serving on the Council, and holding office.[5]

But if in theory some Greeks recognized distinctions between occupations that bear some resemblance to the conventional Chinese schematization of gentlemen, farmers, craftsmen, and merchants, what were the realities? In practice, in most Greek city-states of the classical period there was no soldier class. The fighting was done by the same body of men who had taken the decision, in the Assembly, to go to war and who had elected the generals to lead them. The soldiers accordingly might also perform one or more of the further functions of farmers, craftsmen, traders. Quite a high proportion of the citizens may have owned some land and so in that sense counted as farmers, although the actual agricultural work was mostly done by slaves, and large estates were managed by one or more overseers (who might themselves be slaves).

The main divisions within the state actually observed in Solon's constitution (early sixth century B.C.), for instance, were based on property qualifications. At the top of his four classes came the Five-Hundred-Measure Men (those whose estates produced that amount of grain or the equivalent in oil or wine), below them the Knights (those with three hundred measures), the Teamsters (two hundred), and the Laborers. The last were excluded from holding office, but they were able to attend and vote in the Assemblies and Dicasteries (law courts). These were fundamental rights. In the Dicasteries large groups of citizens (up to 5,001) sat in judgment in both civil and criminal cases, combining the roles of both judge and jury. That is, they were responsible for deciding issues of law, or guilt and innocence and for passing sentence.[6]

Wealth, then, was more important than occupation. It was chiefly on the basis of the way rich and poor were treated that oligarchies were differentiated from democracies. The principle invoked in the latter was that all citizens are equal because they are equal in respect to free birth, and so all—rich and poor alike—should have equal access to the magistracies as well as to the Assemblies and Dicasteries. The oligarchs countered that men are unequal in wealth and that only those who could meet certain property qualifications should be entitled to hold office or even to participate in the political process. As several commentators pointed out, and not just those hostile to democracy, the rule of the many often corresponded to rule by the poor, whereas the rule of the few generally meant rule by the wealthy.[7]

That raises the question of the sources of wealth and the stability of the distinction between rich and poor. The most approved source of wealth was land. But fortunes were made in trade and even in manufacturing. A Greek could even make tidy sums as a speechwriter

(Lysias, Demosthenes), as a sophist, or teacher (Protagoras, Hippias, and as a doctor (though that is a phenomenon of the Hellenistic rather than of the classical period). Some of those who became famous, or even notorious, for their new wealth, in Athens and elsewhere, were not citizens but resident aliens (metics) and so did not enjoy any corresponding political leverage. The accusations leveled at nouveaux-riches citizens in political propaganda, however, indicate that the barriers to entering the ranks of the very wealthy were surmountable.

Something of a stigma attached to engaging in certain activities, but we must be careful. First, we should distinguish between engaging in an activity full-time, to earn a living, and doing so only occasionally. Second, in the vocabulary of political denigration, labels were often attached to individuals on the basis of the activities from which they made their money, whether or not they engaged in them themselves. It was one thing to engage in an activity such as flute-playing or sculpting or even cobbling occasionally and quite another to do so full-time and for a living. Of course, there are significant differences between the three activities just mentioned. Learning to play the flute was part of an ordinary education, so a man was expected to know how to do so. But that was not true of sculpting, which (like doctoring) needs not just practice but some special training. As for cobbling, someone would make shoes only if he was so poor that he could not buy a pair (we may leave aside the case of Hippias, who, we are told, appeared at Olympia dressed entirely in clothes that he had made himself—and that went for his shoes as well—but that was clearly, as we also learn, just to show off).[8]

Those differences aside, the distinction between part-time engagement in an activity and full-time commitment to it was important. Only the latter was frowned on. The term "sculptor" could be applied to either—something to bear in mind when reviewing the evidence for the reported occupations of Greek philosophers or scientists or their forebears. Pythagoras's father, for instance, was a gem engraver, and Socrates' was a sculptor or stonemason.[9] By themselves, such terms do not distinguish between the occasional and the full-time activity, and we need to be on our guard. The insinuation that someone had to earn a living by a craft—the more manual or (as they said) banausic, the better—was one of the commonest ways of scoring points off opponents, whether in comedy, in rhetoric (forensic speeches especially), or in philosophical polemic. When the politician ("demagogue") Cleon is called the tanner, what this means is not that he did any tanning himself but that he inherited his father's workshop of slave tanners.

The final topic we need to consider in these preliminary remarks on Greek social structures concerns the role of education in general and of **literacy** in particular. How far is there any Greek parallel to the Chinese phenomenon of a clearly marked literate elite, access to which depended on a high level of specialized education? The question of levels of literacy in ancient Greece at different periods is extremely controversial.[10] If the extent of minimal competence in reading or writing is difficult to estimate, that of positive fluency is harder still. If we limit ourselves to some very broad generalizations, however, we can make one positive and one negative point.

The positive point is that the elementary education that all citizens were supposed to receive in classical Greece included learning their letters. Some of the institutions of the Athenian democracy depended, in principle, on a minimal literacy. Laws and decrees, for example, were recorded on tablets set up in public that served a publicizing function, even if only a small proportion of those who saw them could actually read them. Again, the institution of ostracism (temporarily getting rid of a political leader considered disruptive) depended, in theory, on each citizen being able to write, on a sherd, the name of the person he wanted to remove from Athens. Yet the archaeological evidence from the first half of the fifth century B.C. shows that corners were sometimes cut. A hoard of 191 sherds, all inscribed with the name Themistocles, written in some fourteen hands, reveals that some were prepared to try to save their fellow citizens the trouble of writing the name themselves.[11]

Yet the positive point remains: by the fifth century B.C. all citizens were supposed to have had some schooling that included reading and writing. The converse negative point is that access to full literacy was not restricted within the citizen body to any particular group. There was no scribal class—even though in Hellenistic Egypt, for instance, Greeks acted as scribes, in the sense of letter writers, for their communities. Degrees of literacy certainly varied, although most exposure to works of high literature would have been mediated in the oral mode. At the same time, books, that is, papyrus rolls, became increasingly available from the fifth century B.C. Anyone could buy Anaxagoras's treatise for a drachma—the equivalent of a day's wage for a skilled laborer—in the market at Athens. Private individuals, as well as philosophical and medical schools, began to accumulate libraries.[12]

Some of the technical treatises, such as the medical works that now form part of what we call the Hippocratic Corpus, may always have been by authors who remained unnamed (even if they sometimes refer to themselves in the first-person singular), although these were not handed on by lineages in the manner described for China. But most of the contents of the early Greek libraries would have been works by identifiable writers, poets, historians, and orators, as well as philosophers.

Who exactly was allowed access to the famous library founded by the Ptolemies at Alexandria is not clear, but it was not for the exclusive use of palace officials. In general, what passed for a literate elite in Greece looks very different from the Chinese counterpart. Education in Greece did not serve to maintain the status of literate families: nor was denial of access to the educational system a notable means of exclusion. There was no such control.

Literacy was important, to be sure, as a way of discriminating among those who practiced as healers or as builders—for instance, between those who claimed higher social prestige and the rest. But the key point is that whereas elementary education was regular and conformed to stable patterns, what passed for higher education, once we can begin to use such a term, from the fifth century B.C. on, was quite unstandardized.

These social structures were not invariant over time. The principal differences that mark the Hellenistic period relate to the changed political circumstances of Greek city-states: their loss of independence. Once first Alexander and his successors and then the Romans held the reins of political power, that had devastating effects, and not just on self-esteem. The Athenians had prided themselves on their total control of their political destinies through the exercise of freedom of speech in their Councils and Assemblies. But when political autonomy no longer existed—or only in a severely attenuated form—free speech was a pale shadow of its former self. It was no longer the norm in political decision making, even if it continued to thrive in competitions of rhetorical display and in the debates of the philosophical and medical schools. With the transfer of political power to Rome, the most the Greeks could hope for was to retain some of their erstwhile intellectual and cultural prestige—which they did, at least for a time. What finally sealed the fate of many key institutions of pagan culture was the conversion of the Roman empire to Christianity, bringing more overwhelming changes both in values and in social structures than the Mediterranean world had experienced previously: but they are beyond the brief of this investigation.

We may turn now to the concrete evidence for our first principal problem, the **social origins** of philosophers and scientists. As with China, our sources have limitations. We may suspect that some authors are indulging in gossip or abuse. We have no way to check how representative our cases are: so we have to add the risks of this invisible bias to the visible ones of slander. Yet for all the shortcomings of the evidence, it has one moderately reassuring feature—

namely, that it does not all fit into a single pattern. Our sources do not suggest that all or even most philosophers and scientists were well-to-do, nor that they were all impecunious. Indeed, this remains true, as a broad generalization, from the classical period right down to the late Roman empire. While we will have occasion to distinguish the patterns of available patronage at different times and places, the social origins of philosophers and scientists throughout Greco-Roman antiquity were much more diverse than in China.

At the wealthy and aristocratic end of the spectrum we may begin with Plato. His father and mother claimed descent from Codrus (a legendary king of Athens) and from Solon, respectively. He was related to several leading statesmen and politicians; they included two of the notorious Thirty Tyrants, who carried through the anti-democratic coup in 404 B.C., for Critias was his mother's cousin and Charmides his own uncle. The Platonic *Seventh Letter* was well informed about Athenian affairs and can be used as a source for them even if not written by Plato himself. It explains that as a young man Plato naturally expected to enter public life, but he was bitterly disappointed in the Thirty (although they asked him to join them) and then also in the restored constitutional government (they put Socrates to death).[13] But Plato was evidently rich besides being well connected and clearly originally intended no other career than that of "statesman," that is, participation of some kind in the political affairs of Athens.

Plato's involvement with the tyrants Dionysius I and II of Syracuse was motivated not by any desire to seek out wealthy patrons who would keep him in style but rather by political ambition. He hoped to persuade Dionysius II to put into practice some of his ideas for the best form of government. But as events showed (when Plato was put under house arrest and had to be rescued by an expedition sent by Archytas), that was a grotesque error in judgment, and one that conclusively demonstrates Plato's naïveté in practical politics.

Other wealthy and well-connected philosophers and scientists can be cited from earlier periods and from later. Anaxagoras, in the fifth century B.C., came from a wealthy and noble family of Clazomenae, although he reportedly renounced his patrimony. Empedocles, who refused an offer of kingship in Acragas, came from a wealthy family; his grandfather was a horse breeder.[14]

Much later, in the third century B.C., Archimedes was on easy terms with, and may have been a kinsman of, Hiero, ruler of Syracuse. Stories of his absentmindedness include one about how, when Syracuse fell to the Romans, he was killed by a Roman soldier while he was concentrating on a geometrical problem. But he was not so otherworldly that he did not contribute very considerable technological skills to the defense of his city in its long-drawn-out siege.

Of those active in philosophy and science later still, in the Roman period, Galen (late second century A.D.) came from a wealthy family at Pergamum (his father was a prosperous architect). His contemporary Marcus Aurelius was not just a notable Stoic philosopher but also emperor.

Although the image of the otherworldly philosopher became a commonplace, in some cases it does not fit other evidence. In the fifth century B.C. Melissus of Samos, a follower of Parmenides, was in his hometown a political leader and general of some renown: he defeated the Athenians, no less, in a naval battle in 441.[15] In the fourth century B.C. the important Pythagorean philosopher Archytas of Tarentum, the organizer of Plato's rescue, was a prominent statesman and undefeated general. The active involvement of at least a few philosophers in politics did not cease after the classical period. Apart from Marcus Aurelius, another Stoic philosopher, Seneca, held high office, although his eventual fate (Nero compelled him to commit suicide) is a reminder that cultivating the favor of autocrats was, as in China, a risky business.

At one end of the spectrum are Greek philosophers and scientists who came from families as wealthy and as aristocratic as any. Let us turn now to the other end—and come back later to the middle range. At the opposite end from the aristocrats there are a few philosophers who came from slave families. The most famous was the Stoic philosopher Epictetus, in the second century A.D. But there were evidently one or two others, particularly among the followers of Diogenes the Cynic. In the third century B.C., for instance, Bion of Borysthenes made some play of his humble background. He was probably exaggerating for effect when he claimed that his father was a freedman who sold salt fish and that his mother came from a brothel. When his father was caught out in tax evasion, the whole family was sold into slavery. Bion was bought by an orator who, however, upon his death left him his estate.[16]

The extent of slave participation in such activities as doctoring and engineering is controversial. Plato refers to slave doctors in the *Laws*, but his aim there is to emphasize how differently a true doctor will behave (he will educate his patients); he is obviously not reporting actual practice.[17] The healers we hear about, who included not just "doctors" (*iatroi*) but also "root cutters," "drug sellers," and "midwives," probably included individuals from every stratum of society, not excluding slaves. Evidence of slaves among the skilled workers in several arts and crafts comes from a variety of sources. The extant financial accounts of the building of the Parthenon show that citizens, resident aliens, and slaves worked side by side in approximately equal numbers. Evidence of the prices paid for slaves at sales shows that those with special skills fetched more, and they might include doctors as well as masons or flute players. Given that writers in other fields were slaves (starting with Aesop in the sixth century B.C.), there is no reason to rule out the possibility of slaves composing other types of work, even if we cannot positively identify any extant medical treatise (for example) as such.

Leaving aside now the specific question of slavery, we can cite a fair number of philosophers and others from comparatively poor families. Some are said to have *become* poor, for example because they renounced an inheritance. But that does not seem to have been why the Stoic philosopher Cleanthes or the mathematician Eudoxus (who was responsible for the first fully worked-out geometric model of the heavens) was poor. Cleanthes was a boxer who worked nights drawing water and grinding grain—to pay for the lectures he attended. Eudoxus originally depended on a doctor called Theomedon for his upkeep ("it was said" he was Theomedon's lover).[18]

Eudoxus and Cleanthes happen to belong to the late fourth and early third centuries B.C., but such cases were not confined to that period. They figure in our sources partly because they typify those who overcame hardship to become learned. A certain outlay was involved: lectures had to be paid for, and, in the case of would-be doctors, a long apprenticeship had to be served and paid for. Although that acted as a deterrent, it was evidently not an insuperable one, even for those of quite humble backgrounds and means.

The middle ground we promised to return to comprises those whose families were neither particularly rich nor especially poor—Aristotle's *mesoi*, or "middle class"—although they were, emphatically, not a *class*. Within the ample space that provides, we can locate Aristotle's own father—court physician to Amyntas, king of Macedon—at the richer end. At the poorer was Socrates' father, Sophroniscus, a sculptor or stonemason, and Theophrastus's, a fuller.[19] These labels leave a good deal unclear, although sometimes other evidence confirms that the families were not well-to-do.

For those in the middle range of inherited wealth, practicing as a doctor, architect, sophist, or philosopher may have been at least a useful way of supplementing their livelihood. Famous doctors and sophists were said to have amassed considerable fortunes. Hippias was said to have earned twenty minae from a visit to a single small city in Sicily in the fifth century B.C. (about two thousand times the current daily wage for skilled workers), but Isocrates in the next century brings us down to earth, for he implies that in his day there were plenty of minor sophists in Athens who barely scraped a living.[20]

We do not have hard evidence to say how far most notable philosophers or scientists could have survived without the earnings they derived from being lecturers or teachers, doctors or engineers. But many doctors and engineers, for example, probably could not. Except in the case of the distinctly wealthy (Galen, say), their work in those areas may well have been their principal source of livelihood—and the point can be extended to teachers in general, once they had a reputation as lecturers. Many who practiced those callings stressed, to be sure, that they were not in it for the money. It was best, some said, not to need to take fees for lectures, and some of the medical writers insist that doctors should be prepared on occasion to treat the sick for free.[21] But behind the facade of the completely disinterested search for truth, it is easy to see that livelihoods were often at stake.

Two main conclusions emerge from this part of our inquiry. The first is a negative one: no single pattern appears in the evidence we have for the social backgrounds of Greek philosophers and scientists of one kind or another from the classical period down to the second century A.D. They were certainly not all nobles, nor all the sons of those who practiced crafts, nor yet all from poor families using philosophy or science in a bid to secure a little upward social mobility. They did not all follow their father's calling (whatever that happened to be). That is the exception among the philosophers, although it may be rather the rule among the doctors.

The second conclusion is the corollary of the first. Recruitment into any of these callings depended less on birth or wealth than on personal ambition and determination. We will have more to say on this when we come to the philosophical and medical schools in the next section, but for now may note how uncontrolled access to those callings was. Anyone could set up as a doctor, although it helped to be able to say who had taught you. More easily still, anyone who could muster an audience could lecture. To succeed there depended on the particular personal qualities, often rhetorical ability, you displayed, not on wealth or social status. If success could and did reflect education, that was always informal, never a matter of passing examinations as candidates for Chinese officialdom did sporadically beginning in the Han and regularly from the end of the seventh century, let alone of taking degrees as at European universities since the twelfth.

Although the political circumstances of Greek city-states changed dramatically after Alexander's conquests, the social backgrounds of those who engaged in philosophy and science remained as mixed in the Hellenistic as in the classical period. In one respect, that heterogeneity even increased, in that more contributions came to be made, in philosophy especially, by those of non-Greek descent. The very first reputed natural philosopher of all, Thales, may have been of Phoenician extraction, but there are not many others to put beside him in the classical period. In the Hellenistic, there is the Phoenician Zeno of Citium, founder of the Stoa. The head of the Academy in the late second century B.C. was a Carthaginian, Clitomachus (his real name was Hasdrubal), whose reputation for philosophizing seems to have antedated his becoming literate in Greek.[22] Later still, with the internationalization of the Roman empire, the phenomenon is commoner still. Plotinus, for example, who was responsible for a revival of Platonism in the third century, was Egyptian, and two of his most important followers, Porphyry and Iamblichus, were both Syrians. That adds one more factor to the variety in social backgrounds that emerges as the main finding of this part of our inquiry.

The fact that a fair number of philosophers, doctors, and scientists earned all or part of their **support** from their practical activities or from teaching indicates that there was some scope for potential patrons, who could, if they wished, save them from the need to do so. This takes us to the problem of patronage. This is, to some degree, a definitional question. At what point does mere employment shade into patronage?

Four specific types of case illustrate the range of possibilities. There are obvious parallels with the situation in China. (1) A doctor being paid to set a shoulder or prescribe a drug falls clearly into the employment category. (2) When a doctor was paid a retainer to be a public physician in a particular city-state, he was accountable to the body that appointed him (the Assembly or Council or some committee delegated to do so), and he had certain contractual obligations for the limited term of his appointment. This retainer was to ensure his availability, not to pay for the treatment of particular cases, for which there would be additional fees.[23] (3) A court physician (such as Aristotle's father) might in certain respects be in a similar position to a public doctor. He would be retained to look after the royal household and might receive a stipend or his upkeep, or both. But his position was less likely to carry specific contractual obligations, might well be for an indefinite period, and depended on the personal favor of the king or ruler. (4) A king, ruler, or wealthy individual could decide to support someone—again, with or without a stipend—specifically to release him from some or all of his usual tasks or duties. The person so supported might be given carte blanche to do whatever he liked, although he would normally be expected to devote himself to some pursuit that would redound to the benefactor's glory or prestige.

This last is the clearest instance of patronage, but the third and even the second cases share the element of a retainer, over and above payment for a direct service. In the original, technical Roman sense, what was in it for the *patronus* was the political support of his *clientes*, although that could be rather a vague matter of general prestige, confirming that the *patronus* had a considerable following. The Roman *patronus* expected loyal attendance from his *clientes*.[24] But the degree to which someone who lived at the court of a ruler was committed to do so might vary. The degree of independence from the ruler would reflect, among other things, whether the beneficiary's services were generally marketable: in that respect, doctors were appreciably less dependent on rulers or rich individuals than, say, lyric poets.

The chief questions here are, first, the extent to which the courts of rulers, or the entourages of the rich modeled on them, were the preferred places of work or centers of attraction for philosophers and scientists and, second, the extent to which powerful and rich individuals positively encouraged and supported philosophical and scientific work. In the third century B.C. the first three Ptolemies at Alexandria had an established reputation for attracting scientists, paying them, and generally supporting their research. How well founded is that reputation, how extensive was their support, and what were their motives? Finally, how exceptional were they in this regard?

We may begin with the question of motives. The two key institutions at Alexandria were the Library and the Museum, the former founded by Ptolemy I Soter, the latter maybe also by him but more probably by his successor, Ptolemy II Philadelphus.[25] In both cases they formed part of a general program of cultural activities that aimed to put the newly established city of Alexandria on the cultural map and to add luster to the fame of the Ptolemies themselves. Their support for science or scientists was just one item among many. Literary scholarship and poetry were as important, probably more so. They were keen, too, to attract philosophers. Theophrastus turned down an invitation from Ptolemy I, although Strato accepted one to teach Ptolemy II.[26]

When they attracted the likes of Eratosthenes (from Cyrene), Herophilus (from Chalcedon), and Erasistratus (from Ceos), it was not so much because they were brilliant scientists as because they were brilliant. Eratosthenes was a mathematician and geographer who was famous for giving an estimate of the circumference of the earth, and Herophilus and Erasistratus were anatomists with many discoveries to their credit, notably that of the nervous system. For the Ptolemies' chief purposes—fame and glory—the mode of brilliance was of secondary importance, whether it came from Homeric scholarship (Aristarchus the Grammarian), from lyric poetry (Callimachus), or from anatomical research. Nor were their motives always just those of prestige. Philo of Byzantium reports that they also supported engineers, but that falls into a different category, for their research—into catapults—had military applications.[27]

The chief office that benefited directly from the Ptolemies' support was that of head of the Library (a post held by Eratosthenes toward the end of the third century B.C.), but they also entirely funded the Museum, where a very mixed group of scholars lived and worked together. We cannot confirm that Herophilus and Erasistratus belonged, but their anatomical researches certainly had the backing of the Ptolemies in one gruesome respect. According to Celsus, the kings provided criminals out of prison for them to vivisect.[28] No one could do that without the ruler's approval and authority, and that may also have applied to postmortem dissection of humans, which was also a rarity throughout Greco-Roman antiquity.

We must be wary of assuming that anyone said to be working in Alexandria—even in the reigns of the first three Ptolemies— necessarily received financial aid or other support from them. Was Euclid a beneficiary? We simply do not know. All we know about him is that he worked in Alexandria; the rest is conjecture or pure fable. Then again, what about Ctesibius, engineer, gadget maker, investigator of pneumatics?[29] He did not have to be attracted to Alexandria; he was born there. But the anecdotes in such sources as Vitruvius that refer to Ctesibius's barbershop suggest a humble origin, and although he may have been taken up by the Ptolemies, we cannot confirm that. The contrast with Archimedes' situation is striking. To prove his dictum "Give me a place to stand and I can move the whole earth" he is said to have exhibited the effectiveness of his compound pulleys, by drawing a fully laden ship to himself single-handedly.[30] Whatever the truth of the story, Archimedes would not have needed to exercise himself unduly to solicit an audience with the king of Syracuse (Hiero), for, as already remarked, he was on easy terms with him. Ctesibius, on the other hand, would have found the task of gaining recognition appreciably more difficult.

We may conclude first that the Museum at Alexandria offered some direct financial support for scientists, among others; second, that it would be a mistake to see every scientist working in that city as a recipient; and third, that support, financial or otherwise, was not targeted at *science* but was part of a wider program aimed at cultural aggrandizement. As Philo puts it, these kings were "eager for fame and well disposed to the arts and crafts."[31]

But how far were there other centers of patronage at all comparable to Alexandria? We hear of institutions like those of the Ptolemies elsewhere, though usually on a much reduced scale. We must be careful, first, about "Museums" for that was a term that was sometimes used of philosophical schools. It was applied, for instance, to Plato's Academy. We will be discussing such schools again later, but in the fourth century B.C. they depended not on endowment nor on patronage but on the fees of pupils. Some other museums were more directly comparable with the Alexandrian in that they were founded and maintained by funds set aside by kings or voted by public decree. The Library, too, had its imitators. Other Hellenistic rulers besides the Ptolemies were avid collectors of books, a trend that stimulated some unscrupulous suppliers to meet the demand by forging what purported to be ancient texts. The most substantial rival to Alexandria was the library built by the Attalids in Pergamum.

The patterns of patronage at Alexandria were also imitated at other places and for similar motives, but if we ask what signs there are that particular scientists or philosophers benefited, the answer is disappointing. We hear of plenty of encounters between philosophers, especially, and kings or tyrants. In the anecdotes the tyrants generally put the philosophers in their place, whereas the philosophers, for their part, try to maintain their intellectual superiority and their independence. In such confrontations, the denouements sometimes favored the philosopher but sometimes exhibited his discomfiture. Behind the fantasy of such stories, it is clear, first, that philosophers did sometimes attach themselves to the courts of kings and, second, that that could be a risky policy. The danger that Plato was in when he visited the courts of Dionysius I and II of Syracuse has already been mentioned. Clearly one would try to cultivate a tyrant with the reputation of a Dionysius only if one were quite desperate—although that is precisely what we are told about another philosopher who went to Syracuse, one called Aeschines Socraticus.[32]

No doubt some tyrants were less dangerous than Dionysius, but few were entirely reliable. One benign ruler, with whom Aristotle got on excellently, was Hermias of Atarneus. Hermias had previously befriended other members of Plato's Academy, and Aristotle stayed with him for three years, married his niece, and composed a hymn in his honor. The trouble in Hermias's case was not that he was some savage autocrat but simply that he was not very powerful: he was eventually liquidated on the orders of the Persian king. On the other hand, stories of tyrants who were ready to do away with courtiers who fell out of favor are not limited to such an extreme case as Dionysius II. Alexander the Great himself set a pattern; among those he had killed was a kinsman of Aristotle's named Callisthenes.

What was the extent of the support that philosophers enjoyed if they were on good terms with rulers? They sometimes received considerable gifts, as in the cases of Cleanthes and Arcesilaus (who was head of the Academy in the early third century B.C.).[33] But although Cleanthes was not well off, neither he nor Arcesilaus depended on those benefactions financially—at least not by the time they had well-established reputations as teachers and had become heads of their schools. The point is fundamental. From the fourth century B.C. on, most philosophers operated most of the time in Athens, where the successful could live well enough on the proceeds of their earnings as teachers. The principal schools were generally self-financing. No doubt they received gifts and bequests from the well-to-do (the wills of the heads of schools sometimes specify property to be left to the school for the benefit of all its members), but the main income, apart from contributions from the members themselves, came from pupils' fees. Eventually, under Rome, the schools received some official support, though only for the head, or so it seems, and only in the case of the principal schools. That modifies the picture a little, but not substantially. The main funding of the schools must still have come from fees, and, most important, they retained their autonomy—at least until Justinian forbade the teaching of pagan philosophy in the sixth century on the grounds that it threatened Christianity.

So far as philosophy goes, we may conclude that the extent of patronage accepted was limited, and what was on offer was not substantial, regular, or reliable. The situation of doctors and engineers or architects was rather different, and that of those who worked in such areas as optics, harmonics, or astronomy different again. We can be confident that ambitious doctors generally sought the richest and noblest patients they could find. Competition to secure the position of physician to any ruler was often intense (even though winning was a mixed blessing, given the risks attached). It clearly was in the case of the imperial household in Rome in Galen's day. Similarly, we know from Vitruvius that competition for kingly patronage in architecture, and, indeed, for commissions for public buildings in general, was rough. But architects, even successful ones, still worked as engineers, not just designing but building and even repairing catapults and ballistae for the army.

The examples from the "exact sciences" mentioned above bring out the importance of the lack of institutional support. The demand for teachers in optics, harmonics, or even astronomy was always very limited, so the potential scope for patronage was that much greater. Yet in striking contrast to the Chinese imperial astronomical bureau, not even work on the calendar received much official support—or

even recognition—in Greco-Roman antiquity. Meton proposed calendar reforms in Athens around 430 B.C., as did Callippus a hundred
years later. Yet the implementation of their results was halfhearted,
even at Athens, let alone in other Greek city-states.[34] Some astronomical instruments were set up in Alexandria, presumably by the
Ptolemies. But the only general source of finance for astronomers
was what they could earn as astrologers (which most of them of
course were). Those who worked in harmonics might be practicing
musicians, and those in optics occasionally designed elaborate arrays
of mirrors to amuse or amaze, such as those in Hero's *Catoptrics*. But
none of that amounts to support for fundamental research; indeed,
none of it amounts to institutional support of any kind.

No doubt the situations of philosophers and scientists of different kinds in ancient Greece and China had many similarities.
The problems that doctors faced in building up a practice were, in
the broadest terms, much alike. However, we may suggest three
possible areas of contrast on the basis of this inquiry, each with
important repercussions on the nature of the scientific work done
in these two societies.

1. Compared with their Chinese counterparts, Greek intellectuals
were far more often isolated from the seats of political power. There
were exceptions. Some Greek and Roman philosophers were involved
in politics, holding office or serving as ambassadors (in the fifth
century B.C. the sophists Gorgias and Hippias did so, as did Carneades
and Posidonius in the Hellenistic period). Plato sought to win over
Dionysius II, and his contemporary Isocrates tried—without much
apparent success—to rally a number of kings and tyrants to support
his Panhellenic policy and lead the Greeks against Persia. But the
ambition to advise a ruler, commonplace in China, was exceptional
in the Greek context. Greco-Roman rulers were not famous for gathering intellectuals around them to tell them how to restore order:
that was hardly Seneca's role, nor had it been Anaxagoras's, nor
Aristotle's. Nor did those rulers need to collect intellectuals to

provide them with what would pass as an orthodox cosmology to legitimate their rule. If they had tried, they would in any case have failed, for no Greek cosmology won out against all rivals, not even Aristotle's in its heyday.

2. There is the lack of bureaucratization: there was no institution analogous to the Chinese astronomical bureau. The indifference of the authorities to advances in astronomical knowledge in Greece is striking. Those authorities were eventually concerned with astrology, but only because they saw it as a threat to stable politics. In Rome, the unofficial casting of the horoscope of the emperor or members of his family was an act of high treason.[35]

3. There is the question of educational qualifications in securing an entitlement—whether to teach or to practice and whether as a doctor, astronomer, or philosopher. There were no such formal qualifications in ancient Greece. Entering the ranks of philosophers or scientists was possible for those of different social backgrounds. It was not necessary to be wealthy or noble: nor did you have to be the son of a doctor to become one. Above all, you did not have to belong to an exclusive, well-marked literate elite. Literacy certainly helped in the battles of prestige that had to be fought. But a high level of scholarly learning was in no sense a sine qua non to become a philosopher or mathematician, let alone a doctor or an architect. By contrast, although Han China had no formal qualifications for specialists, they were eventually expected to memorize the pertinent classics (which often implied membership in a lineage).

The chief preoccupation of the up-and-coming philosopher or scientist, in all periods of Greco-Roman antiquity, was not to find a rich patron but to make a reputation among colleagues, often by confronting them directly in argument. It is this that stimulated, even if it did not dictate, much of the strident adversariality that is such a feature of Greek intellectual exchanges, particularly when these are compared with the discretion and implicitness often exhibited in Chinese styles of persuasion. Our next topic is the modes of operation of Greek philosophers and scientists, their groupings and interrelations, and the solidarity or lack of it between pupils and teachers and among those who considered themselves in some sense colleagues.

Individuals, Groups, Sects, Orthodoxies

We may now raise the companion questions that we posed concerning Chinese collectivities. What did membership in a school or sect mean for a Greek philosopher or scientist, and how common was it? What loyalty or allegiance was expected? How were members recruited; what constraints did they accept; how were deviation and defection viewed? Our evidence is richest for philosophy and medicine in the Hellenistic period, but we may begin with some comments on the period down to Aristotle.

With one principal exception, the so-called Pythagoreans, philosophers before Plato were generally strongly individualistic thinkers, a term we may define negatively, to pick out those who went their own way with no allegiance to any collectivity. The doxographic tradition that stems from Aristotle and Theophrastus pays much attention to teacher-pupil relationships, but many such reports are no more than convenient historiographical fictions.[36] The first three natural philosophers, Thales, Anaximander, and Anaximenes, are sometimes referred to as the Milesian school in modern textbooks, but it is easy to see that "school" is inappropriate in two respects. They evidently did not teach anyone else, and although they may have asked similar questions, they did not share any positive doctrines in answer to them.

The Eleatics—Parmenides, Zeno of Elea, and Melissus—had more in common: they were all monists who denied change. Plato tells us that Zeno set out to defend Parmenides from the ridicule of objectors by showing that their positions were "even more absurd."[37]

It is clear, however, that Melissus was in no way inhibited from departing, at points radically, from the teaching of Parmenides—for example, by asserting spatial infinity, which Parmenides had denied.

As for the rest who figure in standard histories of Presocratic philosophy, most were highly individualistic (like Anaxagoras), and some were represented as deliberate loners (like Heraclitus).

The Pythagoreans were an exception, for two main reasons. Aristotle chose to group them together under that title in his reports on their ideas. From the first century B.C. revivers of Pythagoreanism went all out to secure the authority of the founder for their own ideas. In the process they played down any differences between their own and earlier views and even *within* the latter. Stories about the secretiveness of the Pythagorean group may go back to the fourth century B.C.,[38] but they receive much greater emphasis in such later sources as Porphyry and Iamblichus. There they serve an obvious function. They allow those sources to ascribe to Pythagoras himself ideas that are not otherwise attested—for it became possible to explain the lack of attestation by the rule of secrecy.

But we know on Plato's testimony that Pythagoras himself taught a way of life,[39] and Pythagorean groups were actively involved in the politics of several southern Italian cities in the early fifth century B.C. Reconstructing who belonged to these groups and what belonging entailed is again made problematic by the exaggerations in late sources. But the evidence for three of the most prominent philosophers associated with Pythagoras in the late fifth and early fourth centuries B.C., Empedocles, Philolaus, and Archytas, shows that none of them had any difficulty about adopting some fundamental Pythagorean doctrines but not others, nor about introducing important original ideas of their own, nor about publishing them—that is, writing and circulating books under their own name. So much for a supposed Pythagorean injunction against disclosing the secrets of the sect.

The expansion of education in the fifth century B.C. can be linked, in large measure, to the growth of the activities of the sophists. The problems of evaluating their work are notorious—not least that of penetrating the smokescreen produced by Plato's polemic against them.[40] He accused them of being immoral corrupters of the young—just the charge leveled, he thought unfairly, against Socrates. But from the consumers' point of view, it is apparent that they taught a wide range of subjects, including philology, grammar, rhetoric, mythology, music theory, mathematics, astronomy, physics, and, not least, medicine. The lecturers themselves were in no sense an organized group. On the contrary, each was intent on making his own reputation—and on outdoing his rivals. Some, for that pur- pose, developed extravagant styles of showmanship (Hippias, for example), and in some cases extravagance was a matter not just of style but also of the content of their lectures.

Although they were attacked en bloc by Plato and satirized by Aristophanes, they were highly diverse in their interests as well as in their personalities. Although there are stories of moves made against them (as in the case of Socrates), there was no general attempt to control or curb them: stories of the exiling of Protagoras from Athens and the burning of his books are late fabrications. Many of the sophists were very successful, and not just as teachers. We have mentioned that Gorgias and Hippias served as ambassadors for their home cities, and Protagoras was sufficiently well regarded for Pericles to use him as a consultant on the constitution of the colony he founded. In that sophists provided more than just elementary instruction in a variety of subjects, they fulfilled a genuine and growing function. In taking fees for their lectures, they opened up a possible career for teachers. If the going became difficult in one city, it was easy enough, in the classical period, to move to another.

With the founding of Plato's Academy in the fourth century B.C., the institutionalization of higher education began.[41] Three points stand out immediately as fundamental: (1) the Academy was a private foundation (2) motivated in part by political aims but (3) diverse in the interests and views it allowed to be cultivated. What we call the Academy took its name from a grove or park outside Athens where it was located. The park had been a popular meeting place, including for people giving lectures or teaching, before Plato bought property there and chose it as the location for his school. The school itself was an informal association of individuals with varied interests. Whether fees were originally paid is not certain, but in most schools of the Hellenistic period they were, and contributions were probably expected from members for the upkeep of the school. As noted above, it was not until much later, in Roman imperial times, that the state offered support, paying the head of the school a stipend. This was, then, a foundation established by a private individual, though a wealthy and well-born one, not by a ruler, nor by the state of Athens.

Second, Plato's ambitions for the school included the training of statesmen. His dream was that philosophers should become kings, or kings philosophers, and in the *Republic* he set out an elaborate curriculum of study, including mathematics and dialectic, for the education of those who were to govern his ideal state: the guardians. How closely he followed it in the Academy is not clear. According to the *Seventh Letter*, it was because he saw the need to implement his own philosophy that he went on his disastrous Sicilian escapades.[42] Many of those who joined the Academy also had less idealistic political ambitions than Plato did. However, it was not Plato's sole aim to train the next group of political leaders for Athens or anywhere else.

Diversity of interest was the third prominent feature from early in the Academy's history, and with it went disagreement, including with Plato himself. Thus some (Eudoxus, Callippus) were far more concerned with mathematics and astronomy than with politics. During the twenty years Aristotle was with Plato at the Academy, he no doubt began to develop his distinctive interests—for example, in zoology—and may already have formulated his principal criticisms of Plato's theory of Forms. Nor was this so exceptional. Neither of Plato's two immediate successors as head of the school, Speusippus and Xenocrates, simply tried to preserve his philosophy intact. Both felt free to diverge from, and to criticize, Plato's teaching.

Similar points apply also to Aristotle's own school, the Lyceum, modeled in part on the Academy, though without any specific political program.[43] Aristotle himself came from Macedonia, and in Athens he was just a resident alien without a citizen's rights. The Lyceum did not originally possess buildings but consisted of a group held together by common interests in teaching and research. Aristotle's successor, Theophrastus, acquired property and extended Aristotle's already considerable library. The success of Theophrastus as a teacher may be judged by the report that some of his lectures were attended by audiences of two thousand. These were probably not all fully committed students, but part of the function of the lectures may well have been to *attract* pupils to the school.

Here, too, as at the Academy, disagreement was a recurrent feature of the interrelations of the members. Both Theophrastus and the next head after him, Strato, diverged from Aristotle's views. Theophrastus probably began to do so while Aristotle was still alive.

By the end of the fourth century B.C., Athens had already come to dominate in philosophy—a position it held for nearly a thousand years. Both the pluralism of what was on offer and the competitiveness of rival schools are remarkable. As to pluralism first, one of Plato's rivals, Isocrates, also taught what he called philosophy.[44] What he meant by that was rather different from what Plato meant—not Platonic dialectic but the skills and wisdom the trained orator shows in discussion, especially of practical affairs. Other brands of philosophy were cultivated too, by Stoics and Epicureans, Cyrenaics, Cynics, Megarians, eventually Skeptics of various persuasions. Some of these groups were named after their founder, some by where they met, some by a substantive or methodological principle they advocated. But the question for us now is how they operated and interacted.

The ambitious who wanted either to learn philosophy or to teach it came to Athens. But aspiring pupils could evidently pick and choose among a wide variety of philosophical systems and teachers. Our sources record stories of students attending the lectures of one teacher and then another, some eventually going on to set up schools of their own. The founder of the Stoa, Zeno of Citium, was by no means exceptional. His teachers included a Cynic (Crates), a Megarian (Stilpo), and the head of the Academy (Polemo). The third head of the Stoa, Chrysippus, was taught by the current head of the Academy (Arcesilaus) but went on to criticize him. He was also taught by the second head of the Stoa, Cleanthes, not that he agreed with all of his views, either.[45]

Two centuries later, in the first century B.C., a similar situation still obtained. Aenesidemus attended the Academy but left to inaugurate a brand of neo-Pyrrhonian Skepticism. In the Academy itself, there was a dispute between Philo and Antiochus concerning the true meaning of Platonism, which led, according to some sources, to Antiochus breaking away in a bid to revive the old Academy.[46]

The picture that emerges is one of a veritable free-for-all, and the question that this immediately raises is, What held any of these groups together? Some of the schools had an institutional basis— buildings, a library, an official or quasi-official head. Outsiders saw members as holding certain views in common and as having some kind of allegiance to the person recognized as founder.[47] These points, however, need qualifying.

Membership of a school usually involved contributing to its upkeep, and pupils (however defined) paid for instruction. But the distinction between a regular pupil paying for regular instruction and just anyone paying to hear a particular public lecture or debate may not have been hard-and-fast. Those associated with a particular school, whether as pupils or not, evidently felt no lifelong commitment to it. They could leave whenever they liked, and many did, whether to join another group, to set one up for themselves, or just to leave philosophy.

When the question of the succession to the headship arose, more was at stake. It seems that heads were mostly elected by current members, although who exactly qualified to vote is unclear. The sole criterion for the new head was certainly not perceived doctrinal purity. We saw that with the successors of Plato and Aristotle. When Chrysippus succeeded Cleanthes as head of the Stoa, there was a rivalry and a debate between Chrysippus (who had not been taught by Zeno himself) and Ariston (who had) as to who should represent Stoic doctrine in its disputes with its main rivals—a debate that Chrysippus won. Ariston and some others are said by Diogenes Laertius to have "differed" from the other Stoics, and it is easy to slip into thinking of them as heterodox. But we have to be careful, for that term may suggest more of a fixed orthodoxy for them to diverge from than really existed. The crucial point is that Chrysippus is recognized to have "differed" (Diogenes uses exactly the same word) both from Zeno and from Cleanthes, whom he seems to have left while Cleanthes was still alive.[48]

Although some allegiance to the founder's views was expected, what those views were was a matter of interpretation, even dispute. There is one principal exception to the rule, namely, the Epicureans. They were much more conservative than most of their rivals in that there was less doctrinal development from the teachings of Epicurus himself, and there were fewer defectors. They valued friendship highly, in practice as well as in theory. With the other groups, however, and with both the Academy and the Stoa in particular, the point stands. What the founder himself stood for could be, and often was, renegotiated from one generation to the next.

Argument and debate were thus essential to the activity of the Greek schools in their competitions with one another both for pupils and for prestige. In those circumstances it helped to have allies, and you relied on your comembers for support. Yet you were not necessarily able to rely on them completely, for the possibility of their defection was always there. Although the actual extent of doctrinal uniformity varied between different contemporary members of a sect (let alone between different generations of those who passed as the same sect), all sects acted as more or less stable, more or less well organized and close-knit *alliances for defensive and offensive argument.*

The Greek schools were there not just, and not even primarily, to hand over a body of teaching, let alone a canon of learned texts, but to attract pupils and to win arguments with their rivals. They needed their rivals, the better to define their own positions by contrast with theirs. The debates were generally bitter and intense—although the arguments were all about truth and happiness and peace of mind and living the philosophical life.

These structural and organizational aspects of the way most Greek philosophizing was conducted help to explain two of its distinctive features: its pluralism and its strident adversariality. When differences of opinion existed, on fundamentals or on matters of detail, they were certainly not minimized but explored and exploited. Every school, every individual, had to be prepared to justify their own position, to give an account of it, *logon didonai,* to withstand challenge— for the challenges would surely come. Nor would you just wait to be attacked, for you would set about undermining the positions of rivals as effectively as you could.

We might imagine that the associations of scientists would be very different from those of the philosophers. But more of the picture just sketched for philosophy applies also to music theorists, say, or to doctors, than we might have expected. We are best informed about medicine, on which we will concentrate here.

Two factors complicate the problem. The first is that while there were rivalries among the *literate* medical writers that are similar to those among philosophers, they were not the *only* rivalries that were important in medicine. The pluralism of Greek medicine includes such other styles of healers as itinerant purifiers and sellers of charms and incantations, root cutters, drug sellers, and midwives, as well as the practitioners of temple medicine.[49] The last certainly had articulate defenders, but many of the other groups were at some disadvantage in the public debates that some of the literate doctors chose as their way of making a name for themselves.

The second complication was that some attempts were made in medicine to create a stronger link between teachers and pupils than was typical in philosophy. The main evidence is the famous Hippocratic *Oath*, which ties the pupil to the master in a quasi-contractual relationship. "I will hand on precepts, lectures, and all other learning to my sons, to those of my master, and to those pupils duly apprenticed and sworn, and to none other."[50] This lays down that the pupil should behave toward the teacher in some sense like a son. Examples of doctors who came from medical families are common enough, although there are also plenty of instances where that is not the case. The realities of the situation, however, were that, in medicine as in philosophy, pupils often behaved in a quite unfilial manner toward their teachers, disagreeing with them and criticizing them directly.

Nor is it clear how widely the Oath was applied in one or another of its divergent forms. Plenty of evidence shows doctors, both Hippocratic writers and others, breaking both the spirit and the letter of the injunctions it contains and getting away with it. Particular doctors or groups of them could try to enforce allegiance to some code of behavior, but enforcement was out of the question. There was no equivalent to England's General Medical Council or the U.S. state boards. The legal sanctions to which doctors qua doctors were liable were minimal in Greece (though not so minimal eventually in Rome).[51]

In the fifth and fourth centuries B.C. medical theorists exhibited as high a degree of individualism as the Presocratic philosophers did.

Some scholars used to argue that there were three main groups of doctors then, Coans, Cnidians, and Sicilians. But there is no justification for treating any of these as a clearly defined school.[52] It is true that Galen refers to three *choroi*—bands—but he may be overreacting to the single reference we have in a Hippocratic work to a now lost treatise called *Cnidian Sentences*. The speculative reconstruction of the teachings of that work—on the basis of the attacks made on it—then became the foundation for the theories and practices of the "Cnidian school."

The evidence that Cos was famous for its doctors is good, and some may have chosen to go there to be taught medicine, just as many went to Athens to learn philosophy. But overwhelming evidence also exists of the diversity and disagreement both among the doctors from Cos and among those associated with Cnidos, or indeed with any other Greek city. This comes not just from the extreme heterogeneity within the treatises that make up our "Hippocratic" Corpus but also from the history of medicine preserved in the Anonymus Londinensis manuscript, which lists some two dozen writers on medical issues before the Hellenistic period and neither gives nor implies any school allegiances.

We should conclude that before the Hellenistic period there were no medical schools or sects in the sense of doctors self-consciously united by adherence to a single set of theories or practices. Medical education may have been concentrated where educated doctors collected, as on Cos. But as Plato makes clear, to train as a doctor you went, for example, to Hippocrates, not to a school in the sense of either a building or a group.[53] You listened to Hippocrates, and you attended him as he worked. You may well have read some of the medical texts that came to be produced in some profusion, although it was not until much later, until after Galen (second century), that anything like a regular medical curriculum was established.[54] In some cases, as the *Oath* shows, some attempt was made to tie the student to his master and to his family; but it is unlikely that swearing such an oath was generally a necessary condition for being taught. Although much later, Galen, in his more rhapsodic moments, talks of the "initiation" into the "mysteries" of nature when he writes in praise of anatomy, we should not let his rhetoric blind us to the actual differences he saw between anatomical dissection and a real-life religious initiation.[55] Formal initiation of any sort seems to have been much less frequent in Greek than in Chinese technical education, especially toward the end of our period.

As in philosophy, the situation in medicine changed in the Hellenistic period, when we begin to have much firmer evidence of the operations of certain medical sects. The three main ones usually identified, on the basis of the accounts in Celsus and Galen especially, were the Dogmatists, Empiricists, and Methodists. Even here, however, caution is necessary.[56]

To start with, there never was a group of medical theorists who called themselves Dogmatists. Outsiders, generally critics, invented a Dogmatic "school" associated with beliefs in the possibility of causal explanations that go beyond what is directly observed and in the value and importance of anatomical dissection. But beyond that point, those labeled Dogmatists disagreed fundamentally on most issues in physiology, pathology, and therapeutics. Herophilus and Erasistratus did so, for instance, although both pass as Dogmatists. In their case, it makes more sense to talk of their followers as Herophileans and Erasistrateans. These groups took their name from the individual they adopted as leader, rather than from a doctrinal principle. We may compare the philosophers who called themselves Epicureans and contrast those who saw themselves as Skeptics.

What it meant to be a Herophilean was to follow all or most of his medical precepts, but as in philosophy, some students defected. The most unified and, in that sense, most sectlike of the three conventional Hellenistic groups was that of the Empiricists, who were linked by a methodology from which they acquired their name. They agreed that speculation about hidden causes was futile and that experience and practice were all that counted. But the founder of the Empiricist school in the mid-third century B.C. was Philinus of Cos, and he started out as a *Herophilean*. Herophileans and Empiricists thereafter remained locked in dispute over several generations, and to some extent their views and interests were defined in opposition to each other's.

The Methodists formed a different type of association again. They were a good deal more radical in their rejection of theorizing than even the Empiricists had been. Indeed, they diverged from all previous Greek medical thought and practice in one fundamental respect: they rejected the idea of disease entities as such. The doctor must treat the whole patient and what they called the common conditions—but these were general states of the patient, not diseases. Even though that rejection of pathological theory united those who passed as Methodists, those in the group still openly expressed plenty of disagreement and criticism, including criticism of those whom they and others considered to be the forerunners and founders of the Method—Themison in the first century B.C. and Thessalus in the next.

As in philosophy, then, the forming of more or less close-knit associations was an increasingly important factor in medicine in the Hellenistic period, but the emphasis should once again be put on those terms "more" and "less." Some of the reasons are similar, too, for besides their teaching function, these medical schools acted as alliances for debating purposes. The success of different groups was judged not just by whether their patients recovered but also by the explanations offered for why the patients were sick, why they had to be treated in the way they were, and why they *had* to recover or not, as the case may be.

It was one thing to *be* a good practitioner, another to be *seen to be*— that is, to gain a reputation: and in gaining a reputation, explaining *why* you were successful in treatments could be a major part of the battle, whether you were attempting to persuade potential clients themselves or other doctors. So arguments and justification took pride of place, even when the claim was that results alone counted.

The Empiricists can be used to illustrate the point, because although they took their stand on actual practice, they used an argument to justify their position. It is not eloquence that makes good doctors, Celsus reports them as saying: "Even students of philosophy would have become the greatest medical practitioners if reasoning could have made them so. But as it is, they have words in plenty, but no knowledge of healing at all."[57] That has a fine pragmatic ring to it. But Galen makes a perceptive comment in *On Sects for Beginners*.[58] Dogmatists, as he there calls them, and Empiricists use the same remedies *in practice* (in this, they are unlike the Methodists). It is just that their *views on correct medical method* differ. In other words, the difference between these two "schools" lies in the justifications that they offered for essentially similar practices.

Greek academies and schools look very different from those that have precipitately been labeled as their Chinese counterparts. Whatever the functions of the guests, or clients, the Chinese rulers gathered round them, the sources do not describe institutions or collegial bodies of any sort, much less alliances held together for argumentative or disputational purposes. Although plenty of criticism in Chinese texts is directed both at individual thinkers and at lineages, often without specifying the target, there was generally strong disapproval of open disputation. Although multiple initiation was not uncommon in China, one had to follow the forms of discipleship, and heterodoxy was a dangerous charge to incur.

The Greek philosophical and medical schools may not have exactly depended on each other for one another's existence, but the rivalry between them, in each type of inquiry, was a basic feature of their raison d'être. Their primary function was not to hand on and preserve a tradition, nor yet to supply what could pass as an official orthodoxy. They served the ambitions of their own members more than of rulers. In both philosophy and medicine those ambitions and rivalries between the schools effectively inhibited the formation of orthodoxy across them, at least until well after the second century A.D. The reasons for Galen's dominance in medicine from about the fourth century are another story, implying a different institutional framework.[59] In philosophy "orthodoxy" took on quite different connotations once it became a matter of Christian belief.

Above all, the teacher-pupil relationship looks very different in China and in Greece. Greek pupils were too intent on making their own reputations to pay much attention to the idea that they needed to give their teachers unswerving allegiance. Rather than devoting their best efforts to defending their masters' positions from attack, they were just as likely to be among the attackers. The well-documented cases of defection are too frequent to be dismissed as exceptions. The way in which they are recorded show that they occasioned no special sense of outrage. This is true not just in philosophy but also in the literate medical schools, although for the nonliterate medical traditions we cannot say the same and may perhaps presume the opposite.

One final example may serve to bring home the point, from a text in Plato (not renowned for supporting subversive attitudes to teachers) and relating to mathematics, a field about which we have been able to say nothing so far, because the evidence is so meager. In the dramatic setting of Plato's *Theaetetus*, Theodorus is a distinguished mathematician from Cyrene, and Theaetetus, his pupil (himself eventually to become a brilliant mathematician), is about sixteen years old. In the way Theaetetus's relationship with his teacher is portrayed, two things stand out. First, he goes beyond Theodorus in exploring the properties of incommensurables. Theodorus had stopped at a certain point, but Theaetetus says that he pursued their classification further. When he does so, it is notable that he asks for Socrates'—not Theaetetus's—advice.

Second, Theaetetus even says that Theodorus is flatly wrong about his own, Theaetetus's, ability. This shows Theaetetus's modesty in a favorable light, from one point of view, for he disclaims being as brilliant as Theodorus suggested. Yet what he says is that Theodorus "turns out to be a false witness."[60] Even within the conventions of a fictional conversation, that is a bold remark for a sixteen-year-old to make about his teacher in his teacher's presence. We may think of that other famous, and characteristically Greek, claim that Aristotle made (at what age we do not know). About to launch into a devastating critique of Plato's theory of Forms in the *Nicomachean Ethics*, he expresses regret because the theory was introduced by friends (*philoi*: he does not say his teacher). Friendship and the truth are both dear, but the truth is to be preferred.[61]

Debates, Lectures, Dialogues, Treatises, Commentaries

Ancient Greek philosophers and scientists used a variety of means to express their ideas and to convey their teachings. These include debates, ranging from informal discussions to formal, sometimes competitive public debates, lectures—again ranging from the impromptu to exhibition performances—and a variety of types of writing, among the more important being dialogues, letters, treatises, and commentaries. What light can a study of the means and the media throw not just on the form but also on the substance of the ideas expressed—and thereby on some of the distinctive characteristics of Greek philosophy and science? What can a study of these questions tell us about the social and intellectual contexts in which philosophy and science were practiced in Greek antiquity?

We may begin by picking up three points from our earlier discussion of literacy. First, instruction in reading and writing was part of the basic education every citizen in the fifth century B.C. supposedly received. How far, second, those citizens achieved fluency in reading is another matter, although, third, they faced no social barrier to doing so: in other words, literacy was not confined to a scribal class or a closed elite. However, even then, when a fair number of books (papyrus rolls) of different types were available, most intellectual exchange was in the oral mode. We will review some of the evidence for actual debates and discussions shortly: but even when books were read, they were usually read out, not studied silently and in seclusion. Indeed, most of the works of high literature were not so much read out as *performed*, and this applies to not just tragedies, comedies, lyric poetry, and epics but also such prose works as the history of Herodotus. One famous text that illustrates what may have been a common practice is the introduction to Plato's *Parmenides*. Zeno of Elea says he has written a book in defense of Parmenides' philosophy. But Socrates does not suggest borrowing a copy to study it on his own. Rather, he asks Zeno to read it to him—and they then discuss its contents. Even a work of the sophistication and complexity of Zeno's treatise, then, was normally read out.[62]

How far, nevertheless, did Greek philosophers and scientists use and depend on the written word? Our evidence comes, of course, from written sources, whether books by the philosophers and scientists or other texts that describe and comment on their work. Yet some prominent Greek thinkers wrote nothing themselves. That applies, for instance, to Thales and Pythagoras in the sixth century B.C., to Socrates in the fifth, to Pyrrho the founder of Skepticism in the late fourth, and to Carneades, head of the Academy in the second—and that is just among the earlier philosophers.

Just how some of these thinkers operated is not at all clear, although most of those mentioned had some effective publicists among their close associates. Socrates and Carneades especially were famous for their skills in dialectical debate, Socrates in purely informal encounters but Carneades also in the context of formal teaching in the Academy.

If some prominent early philosophers wrote no books, those who did outnumbered them. Yet oral teaching and oral exchanges were still very important. Because the evidence for real live debates and lectures usually receives little attention, it is worth examining carefully.

Aristotle devoted two substantial treatises, the *Rhetoric* and the *Topics* (including the *Sophistical Refutations*) to the kinds of argument to be used in public and in private debate, and they offer a convenient place to begin. If we ask, first, why Aristotle should have written such treatises at all, the answers are both complex and revealing. In the *Rhetoric* he no doubt wanted both to outdo his predecessors (the authors of books on rhetoric called *Arts*) and to reinstate rhetoric as a genuine skill (in response to Plato's attacks on it). But the fundamental reason why a knowledge of rhetoric was essential was to carry out that part of the good life that related to the moral and political activities of the citizen.

The three modes of rhetoric he distinguishes are (1) deliberative, (2) forensic, and (3) epideictic. The first relates to persuasion on issues of policy, especially in the Assemblies and Councils, but Aristotle also notes that it is useful for giving advice "in private." Forensic oratory is all about winning lawsuits. Epideictic rhetoric here comprises especially the oratory of praise and blame. Although the classification suggests three main contexts in which rhetorical skill is deployed, the skills in question have a far wider relevance to human experience in general. Rhetoric is, after all, the faculty of discovering the possible means of persuasion on any subject whatsoever. In the opening sentence of the treatise Aristotle calls it a counterpart to dialectic (another skill not confined to any specific field of knowledge). He goes on: "All men in a way have a share in both." And why? "Because all men, up to a point, attempt to criticize and maintain an argument, to defend themselves, and to accuse"—eloquent testimony to Aristotle's perception of the argumentativeness of his fellow Greeks.[63]

The *Topics* is even more directly relevant to the work of the philosopher. Again we are given specific answers as to its usefulness, that is,

(1) for training ("gymnastics"), (2) for dialectical encounters, and (3) for the study of the philosophical branches of knowledge.[64] The last is particularly striking, for Aristotle suggests that the most appropriate way to get to the first principles of any branch of knowledge is through an examination of the generally accepted views (*endoxa*). Whereas a philosophical investigation can be carried out by an individual on his own, Aristotle clearly sees dialectic—a joint endeavor—making an essential contribution.

True dialectic differs from "eristic" or contentious argument, where the aim is simply victory. Eristic arguers are sometimes contrasted with sophistic ones: the former aim at victory; the latter are concerned with reputation and making money. Thus, sophistical refutations are those that are merely apparent, for the sake of scoring points.

This explicit recognition of a whole type of contentious debate in which the participants are concerned simply to win, or seem to win, the argument, is remarkable. Moreover, dialectic as a whole, when collaborative, not competitive, also has a formal structure, defining the proper roles of questioner and answerer. In both types of discussion it is necessary to follow certain precepts of good practice. There is more to good behavior than not being peevish or stubborn, or not admitting what is obvious (to take some of Aristotle's examples); one should also stick to what is relevant and know how to deal with any irrelevancies you encounter.

Like the *Rhetoric*, the *Topics* provides plenty of tips for success, depending on the type of argument you are engaged in. Some tricks that are to be avoided when dealing with a cooperative partner are, it seems, permissible in other contexts. So Aristotle discusses how to avoid begging the question but also how to conceal one's eventual conclusion from your opponent in order to secure the necessary admissions from him first. He also tells us that it is a good idea to raise an objection against yourself—for that makes the answerer less wary.[65]

Aristotle is perfectly clear on the *theoretical* distinctions between eristic debate, training sessions, and the joint inquiry to discover the truth. But it is obvious that applying those distinctions *in practice* poses problems. In training sessions you may find yourself attacking the truth and maintaining falsehoods, or demolishing falsehoods not with truth but with other falsehoods. In that context, one should practice arguing either side of the question, both pro and contra any given thesis, to be well prepared for both the role of questioner and that of answerer. When a session is purely a matter of practice might well be problematic, however. At one point Aristotle contrasts objecting to a thesis and objecting to the person maintaining it. You attack the speaker only when he is not cooperating, but it may become necessary to do just that.

Evidently the dialectician has constantly to exercise his judgment concerning the nature of the debate he is engaged in, on his role, and, more particularly, on that of his partner or opponent. With some people, Aristotle notes, discussion inevitably deteriorates, and "with a man who tries by every possible means to escape the consequences of the argument, it is justified to use any means to obtain your conclusion: it is fair or just (*dikaion*) to do so, even if it is unbecoming (*ouk euschemon*)."[66]

The *Rhetoric* and the *Topics* between them suggest a culture preoccupied by, and highly self-conscious about, argumentative debates of various types, ranging from full-scale speeches to question-and-answer sessions, involving cooperative and competitive partnerships, and not excluding the use of dirty tricks of all kinds—to win the lawsuit, get a policy adopted, or simply achieve victory in abstract argument. But the next question is how far the picture Aristotle presents corresponds to the realities of the situation insofar as they can be assessed independently.

The extensive engagement of many Greeks (male citizens, at least) in the various political and legal institutions of the city-states of the classical period is not in doubt, even though the degree of involvement in states under democratic constitutions was greater than in the oligarchies. The litigiousness of the Athenians in particular was legendary. But how far did that politico-legal situation affect early Greek philosophy and science? How far do the modes of argumentativeness manifest in the *Rhetoric* and *Topics* reflect actual philosophical and scientific inquiry at any stage?

The mode of rhetoric that Aristotle defined rather narrowly as the oratory of praise and blame provides a point of entry. Epideictic speeches, in Aristotle's sense, would include the orations given at funerals to celebrate those who had died in battle, examples of which we have in Thucydides and elsewhere.[67] But "*epideixis*" (plural *epideixeis*) is often used more generally of any type of public lecture: the word itself simply means "exhibition" or "display." Those referred to, and exemplified, in the medical writers were public lectures dealing with many topics that we should think of as philosophical or scientific. Two Hippocratic treatises, *On the Art* and *On the Nature of Man*, dating from the early fourth century B.C., are themselves such exhibition lectures and provide evidence on the genre.[68]

The author of *On the Art*, for instance, opens with an attack on rival exhibition givers.[69] There are some who think they make a display (*epideixis*) of their own research by attacking the arts: people who make an art out of vilifying the arts. In *On the Art* the author sets out to defend medicine; other arts will be the subject of other discussions, he says. At the end he refers to people who display the validity of the art not in speeches but in deeds—this different kind of *epideixis* takes the form of acts. But the author's own defense of medicine is not an exposition of medical practice (on which he is rather superficial); rather, it is a set of philosophical or epistemological arguments justifying the claim that medicine is more than a matter of luck; it is a genuine skill.

On the Nature of Man also opens with criticism of other lecturers, this time theorists who propose monistic element theories very like those we know from independent evidence that both medical writers and philosophers put forward. Some lecturers assert that man is all air or fire or water or earth, others that he consists of one of the humors. But if they agree in being monists, they disagree about which the fundamental element is. They can be seen to be ignoramuses, the author says, among other reasons because it is not always the same man who wins the debate. Even "when the same men debate with each other in front of the same audience," the same speaker "never wins three times in succession, but now one does, now another, now whoever happens to have the glibbest tongue in front of the crowd. Yet it is right for a man who says he has correct knowledge about things to be victorious in the presentation of his argument every time—if he really knows the truth and sets it out correctly."[70]

The expression of that notion, that truth will prevail, is itself a remarkable act of faith, or rather a piece of rhetoric, and there are other signs of a real or pretended naïveté in this author, as when he implies that the mere fact of disagreement among the monists is enough to cast doubt on all versions of such theories. Still, three points of some importance emerge from this text.

First, we have to register that such a topic as the fundamental constitution of the human body was a possible subject for a public lecture. We can confirm the point for other technical subjects, too.

Second, these occasions are evidently competitive, with one lecturer trying to outdo another and all competing to win. In On the Nature of Man the contest is apparently adjudicated by the audience—evidently a group without special competence in the field (no peer evaluation this). We do not know how they chose the winner, by voting or by acclaim, but either is possible. In some contests, in poetry and music, for instance, the winner was decided by a show of hands—by the audience as a whole or by judges chosen for the occasion (perhaps by lottery, as was the case for the judges of the competitions in tragedy in Athens). But Plato also often tells of the way the crowd egged on the performances of sophists with applause.

Then a third point will take us a little further afield. The author of *On the Nature of Man* primarily envisages speeches followed by counterspeeches, but public lectures were also often followed by question-and-answer sessions. Protagoras is one of the sophists whom Plato represents as boasting about his ability in such a context—although Plato shows him to be no match for Socrates.[71] Such sessions continued to be a feature not just of sophistic performances but also of philosophical and medical discussions, giving rise to a genre of writing devoted precisely to medical questions, as in the case of Rufus's extant work with that title from the first century A.D.

Another Hippocratic work, *On Diseases* I, gives a vivid idea of what the doctor might expect. This text opens: "He who wishes to ask questions correctly and to answer the questioner and to debate (*antilegein*) correctly on the subject of healing must bear in mind the following things." The questioners will no doubt include the patients themselves and their relatives and friends. The information the author says is needed, and that he himself aims to provide, concerns not only such matters as which diseases are long-lasting and which fatal but also which parts of the body are "hot," "cold," "dry," or "wet," as well as general topics to do with the nature of medicine, including the arts it resembles or does not resemble (compare the preoccupations in *On the Art*). That he has rival speakers in mind, as well as patients, becomes clear at the end of the opening chapter: "Whatever mistake in these matters anyone makes either in speaking or in asking questions or in answering . . . one may, bearing these things in mind, *attack him in reply* (*antilogiei*, "in debate") in this way."[72]

Once a Hippocratic work is identified as or as like a public lecture, it may be natural to discount its contents as not fully serious, as not representative of the best in Hippocratic medicine. Such a reaction may sometimes be partly justified. There are certainly elements of playfulness in the *epideixis* genre that some authors acknowledge. Gorgias, for instance, draws attention to this at the end of his *epideixis* defending Helen: Aristotle indeed tells us that he suggested destroying an opponent's earnestness with laughter, and his laughter with earnestness.[73] The *epideixis* format also partly served the purposes of publicity, and in that context, some ostentation was usual: a showy lecture enticed pupils to sign up for more extensive, and costly, courses.

On the other hand, too dismissive a reaction would be equally mistaken. We should not underestimate the roles that *epideixeis* and debates played in the production, dissemination, and evaluation of ideas in all subjects. In what remained still a basically oral culture, the exhibition lecture and the formal debate were among the more important public contexts in which new ideas were transmitted, developed, and discussed. As for what may seem the shortcomings of those formats, many of these were shared, if to a lesser degree, by other genres. It was not only exhibition lecturers who overstated their cases or who maintained their theses in the teeth of every conceivable objection. Nor were they the only authors who adapted what they had to say to the particular audience addressed. Aristotle, as we saw, insists on that as a skill that both the orator and the dialectician should display.

We may now take stock of some preliminary findings on argument and debate thus far, although these will need qualifying once we have extended the range of our discussion to include such genres as the treatise and the commentary. We started with the observation that both rhetoric and dialectic were highly cultivated in classical Greece, where they came to be the subjects of self-conscious systematic analysis. In the social and political environment in which Greek citizens operated, some rhetorical and debating skills were more than useful: in certain circumstances they were essential, when speaking in the Assembly, for example, or representing oneself in a lawsuit (as people did, although from the fourth century B.C. they could hire others to write their speeches).

Those who *taught* rhetoric and dialectic had anything but a marginal role to fulfill. Plato did his best to marginalize those he dismisses as sophists. But we should not accept his evaluation of them, for their success was often considerable: nor should we lose sight of the fact that Aristotle saw fit to compose a major work on rhetoric. In later times, indeed, professors of rhetoric were comparable in prestige—and in pay—to professors of philosophy.

The repercussions on philosophy and science of a deep-rooted preoccupation with competitive debate were not limited to minor stylistic features in modes of presentation. They extended also to certain recurrent, and at points dominant, traits in the styles of inquiry themselves. The primary point relates to adversariality. For many, the establishing of a philosophical or scientific doctrine proceeded essentially by way of defeating the opposition.

A famous text in Aristotle testifies to the general point: "We are all in the habit of directing the inquiry to our opponent in argument, rather than to the subject matter in question." It is remarkable, first, that this passage comes not in the *Rhetoric* nor in the *Topics* but in his cosmological work, *On the Heavens*. Second, Aristotle makes the observation about other theorists. He is criticizing Thales' view that the earth remains at rest because it floats on water—which does not take into account certain obvious objections. "These theorists," Aristotle comments, "pursued the difficulty up to a point, but not as far as they might have." When he then says, "We are all in the habit of directing the inquiry to our opponent in argument," he is himself doing that very thing—though that is not to say he ignored the relevance to himself of the point he is making. Third, the continuation of the passage is revealing, too. "A man will pursue a question in his own mind no further than to the point at which he finds nothing to say against his own arguments." He means that a thorough knowledge of the subject matter is needed, but again this throws light on

Aristotle at work in his own scientific investigations. Having said that to focus too exclusively on opponents is wrong, he here seems to suggest that in examining your own theories, you need all the help you can get: it is difficult to carry through the critical scrutiny of your own ideas by yourself.[74]

Let us now suggest more speculatively a point to which we will return in Chapter 4. The very adversariality of Greek modes of inquiry seems to affect also the *contents* of theories in this way. It favored systematically exploring the arguments on both sides of fundamental questions (they came to be known as arguments *in utramque partem*—"on either side of a case"). That may well have contributed to a readiness not merely to air but to maintain the contradictory of what might pass as a commonsensical view. It is striking that on just about every fundamental cosmological issue, antithetical views are proposed, and not merely hypothetically but in all seriousness. That includes the theses that the cosmos is one and that there are many of them, that it is finite and again infinite, that it was created and that it is eternal, that matter, space, and time are all atomic and that they are all continua—the list can be extended almost indefinitely. It includes some highly counterintuitive positions, most notably Parmenides' denial of change and Heraclitus's of rest. Realizing that for every major dogmatic thesis the antithesis had also been proposed, the Hellenistic Skeptics make this the cornerstone of their recommendation to avoid all such dogmatizing and suspend judgment. Yet that, too, was a philosophy born out of confrontation with the opposition.

So much by way of some preliminary conclusions. Before turning to the treatise and the commentary, let us glance at the two most substantial and impressive philosophical oeuvres of the fourth century B.C., those of Plato and Aristotle.

With the exception of some doubtfully authentic letters, all of Plato's works take the form of dialogues. These vary in how close they stay to the style of live conversation, the earlier works being closer than the later. Plato stuck with the dialogue form, however; whatever its drawbacks, it secured one fundamental and obvious philosophical advantage. Because the author never speaks in his own voice, the reader has to reflect on the pragmatics of the communicative acts that constitute the dialogue itself. Would Socrates have put the point differently to a different speaker? And what about Plato? Because Socrates always converses with individuals who have their own distinctive views, assumptions, and levels of comprehension, that inevitably underlines the interactive nature of their exchanges. We cannot talk with Plato, but his writings invite, even force, philosophical ruminations similar to those they represent. Thus the form of his writing affects its philosophical content. The preferred mode of philosophizing was essentially dialectical, the presentation of points of view in interaction, sometimes in balanced pairing of speech and counterspeech, more often in questions and answers.

Aristotle also wrote dialogues—in direct imitation of Plato—although only a few paltry fragments of that part of his work survive. We are used to speaking of the extant Aristotelian Corpus as a set of philosophical and scientific treatises, but more strictly they are *lectures*, or his notes for them. The title of the *Physics* in Greek is *Phusike akroasis*, where the translation of *akroasis* is "lecture" or "hearing." These lectures do not take the form of *epideixeis*. Aristotle pays little attention to style, although in places they have been very carefully constructed. If at other points his writing becomes telegraphic and takes the form of a sequence of elliptical notes, that reminds us that Aristotle's oral exposition would have elaborated the argument.

A remark in the *Rhetoric* provides a finely balanced view. In every kind of instruction, he says, there is some slight necessity to pay attention to style, for the way one speaks makes a difference to clarity. The importance is not very great, however, and these are matters of appearance and relative to the hearer (*akroates*), which is why no one teaches geometry in that way (that is, with great attention to style).[75]

The question of how far Aristotle considered his own investigations like geometry remains open. But the text suggests that despite the unpolished character of the extant writings, he was conscious that style contributes something to the effectiveness of all teaching. More important for our concerns, true to the lessons of his own Rhetoric, he is aware that teaching (like public speaking) is interactive. Taking into account points that are "relative to the hearer" serves to remind us that auditors were present, and therefore of the possible influence that their expectations and experience may have had on the philosophical and scientific instruction they received.

We have so far concentrated on the evidence for debates, dialogues, and lectures of one type or another. But treatises consisting in systematic expositions of particular subject matter that take none of these forms were also produced in large numbers in the classical period.[76] Their origins may even go back to the mid-sixth century, although we are in the dark about both the range and the format of the first prose compositions of Anaximander and Anaximenes. By the mid-fifth century B.C., writing on a number of technical subjects is attested, including architecture, agriculture, mathematics, astronomy, music, and medicine. The range of subjects covered by Aristotle's associates in the Lyceum in the fourth century B.C. is considerable. We have works by Theophrastus on botany and mineralogy, for instance, though much else is lost, such as Strato's treatises on physics, Dicaearchus's on geography, and many historical studies of earlier thought. But it was not just the Lyceum that produced technical work; in many scientific subjects, Alexandria came to surpass Athens.

We may concentrate first on the so-called Hippocratic Corpus and then on the group of works on mathematics and the exact sciences that are ascribed to Euclid.

The variety of types of writing in the extant Hippocratic Corpus already attracted Galen's attention when he distinguished works for a specialized readership from those for a wider public, and again *hypomnemata* (clinical notes) from *syngrammata* (compositions of a more connected kind).[77] We have already discussed the exhibition lecture, *epideixis*. But it is not as if the rest of the Corpus is uniform in style or in subject matter. We have works dealing with surgery, embryology, gynecology, and general pathology. The *Epidemics* records particular case histories in detail, interspersed with general accounts of the climatic and other conditions associated with particular outbreaks of diseases. Then, too, we have a number of writings dealing with medical ethics and etiquette, although these are mostly among the latest works—from the third century B.C.

The greater part of all this writing is devoted to the systematic presentation of what their authors offered as useful data. The style is often, though not exclusively, impersonal; the material is set out with little or no attention to literary elegance. Throughout large stretches of many works there are few, if any, signs of the dialectical and rhetorical framework that we see so often elsewhere in the products of Greek intellectual life.

This suggests an important reservation to our general characterization of Greek philosophical and scientific output, but it is itself subject to two qualifications. The first relates to the polemical intentions that are still discernible, even if they are not stridently expressed. Hardly a treatise in the Corpus is entirely free from implicit or explicit criticisms of other doctors. Even when the style is impersonal and the subject matter technical, the fundamental competitiveness of Greek medical practice is evident.

For example, the surgical treatises are mostly taken up with a plain description of what happens in dislocations and fractures of various kinds, of the prospects for treatment, including advice about which kinds of case are hopeless. The writers, however, also engage in sharp criticisms of faulty diagnoses and incorrect treatments. The opening chapter of *On Joints* seems to envisage the kind of open discussion referred to in *On Diseases* I. The writer claims that forward dislocation of the shoulder does not occur (even though many people think it does); he points especially to cases where the flesh of the joint and arm is wasted and the head of the humerus projects forward. In one such case, he says, "I myself got into disrepute both with practitioners *and with laymen* by denying that this appearance was a dislocation. I seemed to them the only person ignorant of what the others recognized."[78] He explains how hard he found it to get them to understand the true nature of the injury. If this text is anything to go by, the lack of references in some of the technical treatises to discussion or debate, including that between practitioners and laypersons, does not mean that such arguments did not occur.

The second qualification we need to make relates to the use of medical works in teaching. None of the Hippocratic treatises is explicitly addressed "to Beginners," as some of Galen's were. But that would not stop them from being used in instruction. If so, how were they used? Some of the major classics of Greek literature were (in part) committed to memory in schools. But what we have in such Hippocratic writings as the *Aphorisms* is rather just a set of heads or dicta that the teacher would elaborate orally. These texts were aides-memoire, but supplements to instruction, not the core instruction itself. For that, in many cases, the oral would be as important as the written mode.

The corpus of work written by or ascribed to Euclid represents one extreme end of the spectrum from the point of view of impersonality. Euclid himself does not just not obtrude: he is invisible, and his life and character are unknown quantities. Nothing, we might say, could be further from rhetoric than mathematics. Precisely. The mathematicians cultivated a style of argument as different as could be from those of the orators, not just persuasive or plausible but demonstrating incontrovertible conclusions by deduction from self-evident axioms. That was certainly the aim, and in practice, in Euclid's own writings, the concentration on demonstration is such that all other aspects of the inquiry, including how the results were discovered in the first place, are ignored.

The conception of axiomatic-deductive demonstration is one of the most striking intellectual products of Greek antiquity. Yet rhetoric and dialectic may, paradoxically, still be present in the background to the development of the quest for certainty, though as negative models. If the notion of mathematicians indulging in rhetoric of the kind used in Greek law courts would be bizarre, so, too, would be the picture of mathematicians engaged exclusively in demonstration—and yet that is the picture that the extant treatises normally present. If we go no further than the question of heuristics and ask why it figures so rarely in our sources (with the one notable exception of Archimedes' *Method*), then part of the answer may lie in the ambition to maximize the claims to incontrovertibility. That was the basis for mathematics' prestige: it was in that regard that it could and did present itself as an altogether superior type of knowledge.

The final genre for us to consider is the commentary, an increasingly important medium of communication from the Hellenistic period on, even if we can trace some of its origins back to the fifth century B.C.[79] Scholarly exegesis of Homer began in the classical period, although whether such interpretations originally took the form of sustained line-by-line analyses is open to doubt. Aristotle, however, remarks in his *Metaphysics* that old Homeric scholars noted small resemblances but overlooked major ones—which may suggest that their work was detailed enough to descend to trivia.[80]

Medical, mathematical, and philosophical texts as well as literary ones came to be the topics of a more and more massive exegetical effort. They include the Hippocratic Corpus, Euclid's *Elements*, and works in astronomy, music theory, and philosophy itself. The works of Aristotle especially were the subject of detailed commentaries, by Alexander of Aphrodisias (in the second century A.D.) and Simplicius and John Philoponus (in the sixth), among many others.

The writings of the heroic figures of the past came to be treated with a respect that borders on awe. If classicism is (1) the study of the past through texts in preference to the direct study of what those texts themselves relate to and (2) the belief that the truth has all been discovered before and merely has to be recovered and preserved, then both phenomena are on the increase from the Hellenistic period onward. Yet we have to enter two important reservations. First, the attitude to the works of those past heroes was never universally and uniformly adulatory. Second, even when commentary was the chosen vehicle of presentation, that did not preclude originality, and not just originality in the exegetical comments on the text in question.

Thus on the first point, criticism of Hippocrates began in the Hellenistic period and continued right down to Galen. His older contemporary the Methodist doctor Soranus cites Hippocrates more often to disagree than to agree with. In mathematics Pappus, at the end of the third century A.D., was criticizing Archimedes, no less, and there are plenty of criticisms of Aristotle even in the authority-oriented writings of Simplicius and Philoponus.

On the question of originality, it is true that many late ancient writers presented what they were doing as preserving past knowledge and nothing more, merely clarifying a few points for the benefit of the dimmer wits of their own day. But we should be wary of that rhetoric. Galen says he takes Hippocrates as his guide, but he has added much to what can be found in the Hippocratic Corpus.[81] Pappus's commentaries and his *Mathematical Collection* contain some outstandingly original work and reveal that he was in touch with a circle of mathematicians who debated one another's work as well as how to interpret past authorities.[82] Simplicius and Philoponus made original contributions to the philosophy of space and time, and Philoponus also to the study of impetus, although the extent of his debts there to such earlier theorists as Strato and Hipparchus is controversial.[83] In all these cases, the commentary provides the medium, or one of them, for important new work.

The point is not that the rise of the commentary made no difference. It clearly did, for both formal reasons and substantial ones. Thus

(1) the focus was now sometimes on points of language, not just points of content. Galen, for example, moves easily between comments on medical practice and philological observations on Hippocrates. (2) The obsession with claiming originality, seen in many works of the classical period, declines. (3) Sometimes the authority of the past replaces evidence or argument as the main justification for maintaining a view—even though Galen, for one, explicitly disavows such a move, saying that the authority of Hippocrates by itself was not enough.

All three points are important, but we should not ignore two others. First, the question of whom to follow remained disputed. Down to the fourth century in all fields, and in most well beyond, there was no orthodoxy. Second, on the crucial issue for our concerns, the rise of the commentary and of classicism did not mean an end to adversariality.

Some commentators did make a notable effort to reconcile authorities from the past, as when Simplicius brought Plato and Aristotle into some kind of concordance, or when Galen did that with Hippocrates and Plato. Yet Galen was well aware of differences between the last two and even exploits them, when it suits him to, in Hippocrates' favor.

When lining up ancient authorities in your own support, the tendency was to argue that they were all really saying the same thing—what you yourself maintained. But in other contexts, adversariality of the most vitriolic kind was still a feature of this writing. This is so especially in the remarks that one commentator makes about his rivals, whether among his own contemporaries or his predecessors. Simplicius was so irate with his contemporary Philoponus that (following standard law-court form) he could not bring himself to name him, but refers to him as "that man." Less histrionically, he was also concerned with outdoing the earlier Aristotelian commentator Alexander of Aphrodisias. From that point of view, the written commentary just adds another medium of polemic to the many that we have already exemplified: the face-to-face dialectical exchanges, the competitive lectures, and the sustained and well-orchestrated debates between rival philosophical and medical sects.

How much light does this rapid survey of genres throw on our original problems: the relationship between the media of Greek philosophy and science and the contents of their theorizing? Our preliminary conclusions stressed the role of rhetoric and dialectic in the social and political experience of the city-states. We identified an influence on philosophy and science with regard to their well-developed adversarial traits, and we suggested further a connection with the readiness to explore and adopt antithetical theses. Many Greeks were prepared to argue both sides of the case not just in the law courts but on fundamental physical and cosmological problems.

The discussion of the more technical writings that take the form of systematic treatises and commentaries suggests some qualifications. They exhibit varying degrees of adversariality. In the technical treatises it is much less overt, and there is less sense of an ongoing debate with rivals. Yet it is also clear that neither the growth of the treatise nor that of the commentary meant an end to adversariality, and the commentary in particular was an effective vehicle of contemporary polemic even when it presented itself as interpreting an ancient text.

The phenomenon of the increasing authority of the past can be exemplified in Greco-Roman antiquity as it can in China. Yet the way that authority operated differs in important respects. The Greek classics came to be revered and much studied as texts, not just as sources of knowledge. Philology exerted a growing influence from the early Hellenistic period, both as a learned discipline in its own right and for what the close study of ancient texts could contribute to philosophy and medicine themselves.[84] That sometimes went with a sense that no important new truths remained to be discovered, that the job of the scholar was recovery and preservation, no more. Yet that did not represent everyone's point of view, in the third century B.C., in the second century A.D., or even in certain cases later. It would be a travesty to interpret Galen or Ptolemy thus, or even Porphyry or Pappus. True, they all looked back; yet they did so with a critical eye, and they all exercised their right to intervene in the first person in ongoing debates.

The way Greek classics were used reflected the varying degrees of consensus on what they stood for—between philologists, on the one hand, and philosophers and doctors, say, on the other. Within the growing study of philology, there was general agreement on the questions worth asking about vocabulary, grammar, syntax, style. From those points of view, it did not matter to the scholar whether the text in front of him was Homer or Hippocrates or Euclid. Philological scholarship worked within the framework of a very firm methodology.

But in the disciplines that some of the texts address, other issues were at stake. Here, too, in the Hellenistic period, scholars had a greater sense of how to formulate the key questions. But agreement did not extend to a consensus on the answers to be given, or to anything approaching an orthodoxy on fundamental problems.

For a doctor or a philosopher the choice of a text to comment on could itself be a significant move and a key step in self-definition. Even when the stated aim was the preservation of past truths, that was not just what the aim was. In late antiquity, as in the fifth century B.C., the success of an author depended less on pleasing a powerful patron than on impressing colleagues and potential pupils. The crucial point is that authors recognized that there was room for maneuver with respect to which texts to privilege and how to deploy the authority they represented. The struggle over what Plato or Aristotle stood for bears certain resemblances to the changes and developments in the interpretations of Confucius. But there is this important difference, that in the Greek case, there was little sense that orthodoxy was a desirable goal toward which to work. The lack of consensus among their teachers was something that the students of Simplicius and Philoponus must have realized early in their careers.

In China the memorial to the throne was important as a form for presenting technical and other practical advice to those in authority. Whereas some Greek and Latin treatises, too, were addressed to rulers, more often the targeted audience was a general one— colleagues, if not the "crowd" attending lectures, mentioned in *On the Nature of Man*. That difference reflects the *comparative* unimportance, in Greco-Roman antiquity, of the need to find a kingly patron (corresponding in part to a lack of those willing to play the role of patron as well as king) as compared to the reputations to be made in public confrontations with rivals.

The dominant impression this part of our investigation leaves is the influence, on so much Greek philosophy and science, of the model of the competitive debate in front of a lay audience—debate as competitive lecture, as speech and counterspeech, as question and answer. These are not just the models for types of writing; they were also often the actual formats for the presentation of ideas.

Much Greek philosophy and science thus seems haunted by the law court—by Greek law courts, that is, where there were no specialized judges, no juries limited to a mere dozen people, but where the dicasts could number thousands of ordinary citizens acting as both judge and jury. No one who has a philosophical or scientific idea to propose in any culture can fail to want to make the most of it. But a distinctive Greek feature was the need to win, against all comers, even in science, a zero-sum game in which your winning entails the opposition losing. In Greece, even in philosophy and science, the competitive often dominated over the cooperative and the consensual. It was not the usual style to think of everyone having insights worth preserving, of everyone making a contribution to the truth.

not represent everyone's point of view, in the third century B.C., in the second century A.D., or even in certain cases later. It would be a travesty to interpret Galen or Ptolemy thus, or even Porphyry or Pappus. True, they all looked back; yet they did so with a critical eye, and they all exercised their right to intervene in the first person in ongoing debates.

The way Greek classics were used reflected the varying degrees of consensus on what they stood for—between philologists, on the one hand, and philosophers and doctors, say, on the other. Within the growing study of philology, there was general agreement on the questions worth asking about vocabulary, grammar, syntax, style. From those points of view, it did not matter to the scholar whether the text in front of him was Homer or Hippocrates or Euclid. Philological scholarship worked within the framework of a very firm methodology.

But in the disciplines that some of the texts address, other issues were at stake. Here, too, in the Hellenistic period, scholars had a greater sense of how to formulate the key questions. But agreement did not extend to a consensus on the answers to be given, or to anything approaching an orthodoxy on fundamental problems.

For a doctor or a philosopher the choice of a text to comment on could itself be a significant move and a key step in self-definition. Even when the stated aim was the preservation of past truths, that was not just what the aim was. In late antiquity, as in the fifth century B.C., the success of an author depended less on pleasing a powerful patron than on impressing colleagues and potential pupils. The crucial point is that authors recognized that there was room for maneuver with respect to which texts to privilege and how to deploy the authority they represented. The struggle over what Plato or Aristotle stood for bears certain resemblances to the changes and developments in the interpretations of Confucius. But there is this important difference, that in the Greek case, there was little sense that orthodoxy was a desirable goal toward which to work. The lack of consensus among their teachers was something that the students of Simplicius and Philoponus must have realized early in their careers.

In China the memorial to the throne was important as a form for presenting technical and other practical advice to those in authority. Whereas some Greek and Latin treatises, too, were addressed to rulers, more often the targeted audience was a general one— colleagues, if not the "crowd" attending lectures, mentioned in On the Nature of Man. That difference reflects the comparative unimportance, in Greco-Roman antiquity, of the need to find a kingly patron (corresponding in part to a lack of those willing to play the role of patron as well as king) as compared to the reputations to be made in public confrontations with rivals.

The dominant impression this part of our investigation leaves is the influence, on so much Greek philosophy and science, of the model of the competitive debate in front of a lay audience—debate as competitive lecture, as speech and counterspeech, as question and answer. These are not just the models for types of writing; they were also often the actual formats for the presentation of ideas.

Much Greek philosophy and science thus seems haunted by the law court—by Greek law courts, that is, where there were no specialized judges, no juries limited to a mere dozen people, but where the dicasts could number thousands of ordinary citizens acting as both judge and jury. No one who has a philosophical or scientific idea to propose in any culture can fail to want to make the most of it. But a distinctive Greek feature was the need to win, against all comers, even in science, a zero-sum game in which your winning entails the opposition losing. In Greece, even in philosophy and science, the competitive often dominated over the cooperative and the consensual. It was not the usual style to think of everyone having insights worth preserving, of everyone making a contribution to the truth.

There is a final irony. Even when some criticized the competitiveness of the model that derived from rhetoric in general and from the legal domain in particular, they did so sometimes in the name of a new kind of competitiveness. Plato reinterpreted the goal of rivalry as the truth, which, as he puts it in the *Gorgias*, is never refuted.[85] His alternative to the opportunistic speeches and arguments of ordinary debaters—their kind of logos—was an impersonal, objective truth, expressed in a logos of an entirely different kind, which could claim to defeat all rivals. The Chinese Possessor of the Way did not need to press this point. His roots in a tradition, the descent of his teachings from antiquity, prepared him to champion those teachings as correct; truth was not the issue. But Masters of Truth, in their endlessly confrontational milieu, had reason to make an issue of truth and defend their claims against all comers. These are themes to which we will return in our next chapter, when we explore the characteristic intellectual products of Greek philosophy and science.

4 The Fundamental Issues of Greek Science

The Problem

The central issue that this chapter addresses is the way in which certain *questions* came to be seen as fundamental for much Greek philosophy and science. Why did Greek philosophers and scientists focus so often on the constituent elements of material objects, on their natures, on the imperceptible reality that underlies the appearances? Why was there so much concern for the causes of phenomena and for the representation of the cosmos as an ordered whole? At first sight it might seem absurd to pose that problem. For are not these among the most obvious and unavoidable questions that any philosophy or science must tackle? If so, is it not utterly superfluous to ask why Greek philosophers and scientists did so?

To that, there are three counters. First, these questions were not obvious and explicit at the very beginnings of Greek philosophy and science; we can certainly investigate how they came to *acquire* the central importance that they later possessed.

Second, reflection on early philosophy and science elsewhere—especially in China—confirms that it is perfectly possible to treat quite different concepts as central. The Chinese, as we will see, spoke of phases (hsing), not elements; they had no single concept that corresponds to "nature." According to one view, they were concerned not so much with causes as with correlations and configurations of change. Each of these points will be elaborated and qualified in Chapter 5, but they show well enough, at a first stage of discussion, that there was nothing inevitable about most of the Greek ways of formulating the principal questions. The great advantage of comparative studies, as we have noted, is to reveal the diversity of scientific traditions—provided, of course, that we avoid the trap of treating the concepts in play in other traditions merely as equivalents of the ones we are used to.

Third and most directly, the problems that the ancient Greeks used these concepts to address have not remained constant and do not look the same to us today. Although many Greek terms, as conventionally translated into English, may generate the impression that they persist unchanged over time, that, too, is a trap. Even when the scientists and philosophers who use the concepts today see them as continuous with ancient Greek speculation, the concepts in question have in every case changed their meaning. The fortunes of the term "physics" illustrate the point. The modern word is derived from the Greek phusike, "the study of nature," the emergence of which we will be discussing in a minute. But hardly a single component of Greek phusike survives in what physicists of today would recognize as their subject matter.

Our first study will tackle the question of when the elements of things, or their natures, or the reality that underlies the appearances, became the explicit focus of attention. That historical investigation will throw light on why the problems came to be formulated in the way they were. We will then tackle the Greek preoccupation with causes (p. 158) and their debates about the nature of the explanations to be attempted in different areas of inquiry, in particular about the relationship between "mathematics" and "physics" in that respect

(p. 165). Finally we will consider the chief assumptions the Greeks made when accounting for the cosmos as a whole (p. 174). In each case we will see that there was nothing preordained about the ways they framed the problems. We will endeavor to show what the key concepts and questions owed to, and where they departed from, pre-philosophical thought. More important, we will investigate what they owed to the characteristic features of the intellectual exchanges and of Greek culture more generally that we discussed in Chapter 3.

Why Elements, Why Nature, Why Reality versus Appearances?

By the time of Aristotle, the terminology of "elements" (stoicheia), "nature" (phusis), and "substance or reality" (ousia) was well and truly established and at the center of much philosophizing. But in the earliest extant Greek evidence, none of these terms is commonly used, and none is used in what became the main philosophical senses. That evidence takes the form of literary texts—Homer's epics, Hesiod's didactic poetry, and archaic lyric poetry. If the philosophical senses had just been echoes of well-established usages in ordinary discourse, they would certainly have left a mark somewhere in those early writings.

The term for substance or reality, ousia, is a noun formed from the verb "to be," einai. Homer and Hesiod already drew certain contrasts between how things are and how they appear (phainetai) often enough, though in neither case with the idea of determining the underlying physical causes of the "phenomena" (that is, those appearances). But Hesiod especially made important claims to special knowledge, and this is important. In the Theogony, for instance, he says he has been inspired by the Muses to sing of the genealogies of the gods, and at the end of the Works, he gives instruction on a subject about which "few men know," namely, which days of the month are propitious, and which unpropitious, for which activities (p. 155).[1]

The second of the three terms mentioned, *phusis*, "nature," illustrates the key issue clearly. Neither Homer nor Hesiod had an overarching concept or category that picks out the domain of nature as such. The term *phusis* occurs once in Homer in relation to a plant, where it refers either to its character or (more probably) to the way it grows. On either interpretation, it does not signify nature as a whole, in the sense of what natural science studies. Besides, the plant was a magic one, called *molu*, which, on Hermes' instructions, Odysseus used to defeat the charms of Circe.[2] The notion of regularities in what later would be called natural phenomena is implicit in many early Greek texts. We should certainly not imagine that the ancient Greeks, or anyone else, were of two minds about the need to sow seed if they wanted grain. But there is all the difference in the world between an implicit assumption and an explicit concept.

The term that eventually came to be standard for "element," namely, *stoicheion*, originally referred to what is drawn up in a line or row (*stoichos, steicho*).[3] In some of its earliest attested occurrences it is used of letters, the primary components of words, as opposed to syllables. It is so used, for instance, in prominent passages in Plato's *Cratylus* and *Theaetetus*. When in the latter dialogue he uses it of material elements, he introduces it half apologetically: "the primary elements, as it were, of which we and everything else are composed." The qualification suggests that Plato expected his audience to find the term, in that sense, unfamiliar. Yet when he presents his own element theory, in the *Timaeus*, he did not hesitate to use the term.[4]

The other context in which *stoicheion* may perhaps have had a semi-technical use before Aristotle was mathematics. Aristotle explains its use for certain primary geometrical propositions the proofs of which are implied in the proofs of all or most of the others.[5] Euclid's textbook (traditionally dated around A.D. 300) has the title *Elements*, and we know that it drew extensively on earlier work. Greek historians of mathematics represent many pre-Euclidean mathematicians, going back to Hippocrates of Chios around 430 B.C., as having put together books that these historians duly call *Elements*.[6] We cannot confirm from direct evidence that *Elements* was the title of those books (although Aristoxenus did use it for his musical treatise in the late fourth century B.C.), nor how much earlier than Aristotle the term may have been used by mathematicians in the sense that he explains. He appears to be explaining an existing mathematical usage, not proposing a coinage. While a possible pre-Aristotelian mathematical usage tells us nothing about the terms in which early physical theorists formulated their ideas, we will have more to say about possible interactions between physical and mathematical element theories from the fourth century onward (p. 154).

The upshot of this very rapid philological survey is straightforward and uncontroversial. *Stoicheion*, *phusis*, and *ousia* each have some relationship with some prephilosophical or nonphilosophical usage. These are, however, quite distinct from their chief philosophical senses. Moreover, their appearance in well-attested texts antedating the fourth century B.C. is rare.

So, what concepts did the early Greek philosophers themselves use for the inquiries they engaged in? Aristotle regularly represents them as fumbling toward his own ideas. But his bias should not mislead us. It is particularly important to be on our guard where he claims that they were searching for "elements," or what he calls the material cause. The fifth-century B.C. philosopher Empedocles has a term, *rhizomata* (literally, "roots"), that he applies to earth, water, air, and fire. Aristotle later adopted these as his four primary or simple bodies. Yet even here caution is in order. Empedocles had two other principles, Love and Strife, which he sometimes treated as on par with the roots. Although Aristotle treats Love and Strife as botched attempts to identify the efficient causes of things, as opposed to what he interprets as Empedocles' four material elements, both interpretations go well beyond the texts of Empedocles himself.[7]

A similar objection applies to much of Aristotle's famous report on the person he treats as the founder of the inquiry into the material cause of things, namely, Thales. He begins his discussion: "Most of the first philosophers thought that the principles that take the form of matter are the only principles of all things. That of which all things are made, from which they first come to be, and into which they are ultimately resolved . . . this they say is the element and this the principle of things." He accordingly treats Thales as having suggested that, as Aristotle puts it, the material cause is water.[8]

Yet that is not all there was to Thales' original idea, as other evidence in Aristotle himself makes clear. In *On the Soul*, for instance, he reports that Thales held that "all things are full of gods."[9] Aristotle objects that those who treat the primary element as alive cannot explain what makes an animal an animal. It is very unlikely that Thales' water was inanimate stuff.

Although most of the terms that Aristotle uses of his predecessors did not acquire their philosophical senses until the fourth century B.C., sometimes at the hand of Aristotle himself, that does not apply to the term translated above as "principle," *arche*. The second of the Milesian philosophers, Anaximander, may even have used it, although the point is disputed.[10] He called his principle the Boundless, and, like Thales, he may well have thought of it as alive.

Whether or not Anaximander himself used the term, *arche* had a variety of senses in physical theory before Plato, and that variety is important. The Aristotelian tradition often took *arche* to mean "principle"—and then assimilated it to "element" in the sense of the fundamental constituent of physical bodies. But *arche* has at least two other primary meanings: "rule or command" and "beginning or starting point." The sense "starting point" is pertinent to our immediate concerns for two reasons.

First, "starting point" is evidently the primary sense in Anaxagoras, for example, who compares how things were at the start of the cosmic process with how they are now.[11] It is possible that the earlier Presocratics were chiefly concerned about the origin of things, rather than their present material constitution. Anaximander's Boundless, for instance, seems to have been an answer to the question of what things came from, rather than what they are made of now. Tables and chairs, and men and women, were not, on that view, made of the Boundless, but came, ultimately, from it.

That is only a conjecture. But the second reason why it is important to bear the possibility in mind is that the question of origins can be said to have pre-philosophical Greek antecedents. Hesiod's *Theogony* offers an answer of a kind to that question. It states that first of all "chaos" came to be, chaos in the sense of "yawning gap" rather than disorder.[12] After introducing Earth and Eros, the poem continues with the genealogies of the gods. By contrast, Thales' account started from water, and Anaximander's with the Boundless. Yet they, too, may have answered the same question—What was there in the beginning?—even if they did so in a radically new way. We shall find other examples of the same pattern, where one Greek Master of Truth outbids others either by reformulating the problems or by providing different answers.[13]

If the dangers of accepting the Aristotelian account of his predecessors' concerns are obvious, is there not one basic concept that escapes that stricture—namely, *phusis*, "nature"? Did not all or most of the Presocratic philosophers share an interest in nature and in naturalistic causes or explanations, an interest that would justify Aristotle's frequent references to them as "natural philosophers," *phusikoi?* "Nature" is indeed a more promising general rubric than many, but one that in turn poses problems of interpretation.[14]

First we have to be wary of the doxographers' reports that many of the writings of the Presocratics went under the title "Concerning Nature," *peri phuseos.* They are generally quite indiscriminate in their attributions of that title, even though some fifth-century B.C. works undoubtedly had it.

Some sources refer to the "inquiry concerning nature" *as such* well before Aristotle. For example, the Hippocratic treatise *On Ancient Medicine* associates this inquiry particularly with Empedocles.[15] When Plato attacks certain theorists in the *Laws* for denying the existence of the gods, he represents them as claiming that the primary factors at work in the universe are nature and chance.[16]

The more extensive evidence in another Hippocratic treatise, *On the Sacred Disease*, reveals that in the fifth century B.C., so far from being a merely neutral or descriptive category, the concept of nature already played a distinct polemical role.[17] That medical writer refutes in detail the notion that the sacred disease was, as its name implies, brought about by divine or demonic forces. This view he attributes to groups of people he calls purifiers, whom he also dismisses as "quacks," "charlatans," and "magicians." His own detailed description supports the view that he is referring to epileptic seizures. In his polemic the key concepts that the writer uses are, precisely, "nature" (*phusis*) and two words for "cause" (*aition* and *prophasis*). The gods, the divine, are not responsible (*aitios* could also mean "the guilty party, the person who is to blame"). Because the disease has its own nature, there is no need to attribute it to some divine agency, nor any justification for doing so, nor is the right way of treating it a matter of ritual purifications. We should add, however, that the writer's ideas, both about its causes and its cures, are quite fanciful, even though the types of factors he invokes—the blocking of the veins by phlegm and the control of the patient's diet—are naturalistic ones.

In this way "nature" came to identify the domain over which the philosophers and medical writers claimed to be able to give credible physical explanations. Many of the phenomena in question —earthquakes, lightning and thunder, and eclipses, as well as diseases—had often been considered portentous, frightening, the work of the gods. Treating them as part of nature implied that they were investigable. Indeed, the claim was usually much stronger, namely, that the offered explanations gave the true natural causes.

Nature came to be a focus of attention in the fifth century B.C. in part because that concept enabled a wedge to be driven between, on the one hand, the new-style naturalistic accounts of events and objects and, on the other, the traditional or conventional ideas that those accounts were seeking to replace. The naturalists' mode of explanation was superior—at least so they believed—precisely because it was in terms just of nature. Their explanations eliminated, in theory, the arbitrary, the willful, the arcane, in favor of what was, in principle, regular and observable. Yet the introduction of the concept of nature was not just the outcome of cool intellectual analysis, for those who invoked it did so in a bid to defeat their rivals. The debate was a competition for prestige and, for teachers and doctors alike, sometimes also to secure a livelihood.

Yet it was not as if the naturalists agreed among themselves about the *nature* of nature, any more than they did about many other aspects of cosmology. Indeed, by the end of the fifth century B.C., nature was at the center of an intense debate that went well beyond explaining physical phenomena to contest the very foundations of morality.[18]

The argument was about the boundaries and relationships between nature and what was often construed as its antonym, *nomos*, a term that covered law, custom, and convention. Some held that morality has no natural basis and that *nomoi* are always relative to societies, groups, or individuals. Herodotus illustrates that point of view in a famous story, according to which the Persian king Darius arranged a confrontation between some Greeks and some Indians at his court on the question of how one ought to treat the dead bodies of one's parents. The Indians were just as horrified at the Greek custom of burning the dead as the Greeks were by the Indian practice, reputedly, of eating them.[19] Yet others argued that there are universal—if unwritten—moral precepts. Some sharply distinguished the domain of nature from that of morality; for others, nature sanctions the principle that might is right; for yet others (including both Plato and Aristotle), nature is steeped in positive values.[20]

By the end of the fifth century B.C., the *terms* of fundamental physical and cosmological problems received sharper definitions, although the contents of the solutions offered were as disputed as ever. Aristotle, especially, set to work clarifying the questions, and by redefining them he ensured that his own answers would be superior to those of all his predecessors. In his view, earlier naturalists were confused about the issues that needed explaining. His theory of causes (pp. 158 ff.) offered a new analysis of the four types of question to be investigated, although we have already raised doubts about how well this works as an interpretation of the problems his predecessors were interested in.

Aristotle's own element theory, based on the four simple bodies— earth, water, air, and fire—exercised a dominant influence on later Greek physical speculation. The reasons for that are complex, nor should we exaggerate the extent of its dominance. In part, it succeeded because he constructed such a comprehensive system on the basis of his analysis of the questions to be investigated. In part, too, Aristotle's theory stayed closer to what one can directly observe than did the theories of some of its main rivals (such as atomism).

The Stoics, for instance, who may or may not have had direct access to Aristotle's physical treatises as we know them, adopted a similarly qualitative account. They, too, thought of earth, water, air, and fire as the elements of compound bodies, although they had no place for the fifth element, *aither*, with which Aristotle accounted for the eternal circular movements of the heavenly bodies. Circular motion conflicted with his view that the natural movements of the four simple bodies were either up (like the movement of fire) or down (like earth). The heavens could not be made of those elements unless their movements were unnatural, and that was unthinkable. Moreover, the Stoics analyzed the ultimate principles differently, distinguishing between an active and a passive principle, the one called god or reason, the other matter. Both these principles, however, are corporeal. By making the active principle immanent in the universe, they returned to something like the belief of some Pre-socratics that the universe is alive. Their *pneuma* was the Greek concept closest to the Chinese *ch'i*, simultaneously material and vital.[21]

Among the scientists who offered element theories, qualitative accounts based either on Aristotle or on the Stoics, or on an amalgam of both, were the general rule in the Hellenistic period. The combination of possible Aristotelian and Stoic features in Ptolemy's physics is a matter of some controversy.[22] But Galen leaves us in no doubt that he saw his four-element theory as having the authority not just of Aristotle but also of Plato and, indeed, Hippocrates.[23]

Qualitative theories of matter were far from having it all their own way. The alternatives on offer from the Hellenistic atomists and skeptics are crucial to our understanding of just how far there was agreement at least on the fundamental *questions* that had to be addressed.

From its very beginning, in the theories of Leucippus and Democritus in the fifth century B.C., atomism offered a view that was opposed to continuum theory on a series of basic issues. Matter is not infinitely divisible, but must be imagined as constituted by indi-visible particles, differing only in shape, position, and arrangement and separated by the void. The atoms vary in size, but they are, in principle, invisible. Yet their differing combinations account for all the diversity apparent in the objects we experience. The void guaranteed not only plurality but also movement, for it is that through which the atoms constantly move. But atoms and the void alone are *real*. Such qualities as hot and cold, far from being fundamental, are matters of convention only—*nomoi*—arising from the interactions and combinations and separations of the atoms.

Other doctrines were more or less closely tied to the atomist view of the ultimate reality. First, some atomists held that not just matter but also time and space are made up of indivisibles. Second, against the view of a unique cosmos (whether eternal or created), they believed in an infinite number of worlds, separated from this one in time or space, or both. Third, they rejected final causes, although they were still profoundly concerned with other kinds of cause that picked out the physical interactions of things.

The post-Aristotelian version of atomism, elaborated by Epicurus in the fourth century B.C., provides another clear example of the interactions of physics and morality. Epicurus was so impressed by the deterministic implications of a fully mechanistic universe that, to guarantee free will, he postulated atomic swerves—random movements of atoms that break the chain of cause and effect and that are, by definition, uncaused. Despite this bid to rescue morality from physics, he maintained that physics is essential to morality. Peace of mind and happiness depend, first, on understanding the fundamental truths of the atomic theory, the doctrine that atoms and the void are truly real. Second, you needed to know that even problematic phenomena, such as earthquakes and eclipses, have natural causes. He insisted that if several explanations are possible, all should be kept in play. To choose between them is to fall into speculation. This principle of plural explanations led him to retain some ideas—for example, on eclipses—that specialists had long ago ruled out (p. 161).[24]

Both atomists and continuum theorists agreed on the fundamental importance of elements, even though they resolved the problems so differently. Both philosophies were dogmatic, from the skeptics' point of view, in that they aimed to explain the underlying reality and the hidden causes of things. Skeptic reactions to that whole enterprise took various forms. Some denied that knowledge on such subjects is possible—the position known as negative dogmatism. But in the Hellenistic period another, far more radical version of skepticism emerged, taking the fourth-century B.C. Pyrrho of Elis as inspiration. The Pyrrhonists were not negative dogmatists; instead, they withheld judgment on all the questions about which the negative dogmatists denied that knowledge was possible. The Pyrrhonists maintained that for every argument that might suggest a positive answer to any question to do with the underlying realities, there was an equal and opposite argument to suggest a negative one. Naturally enough, they were in an excellent position to exploit the very controversies we have just been talking about.[25]

One example suffices to show both how they trumped their dogmatist opponents and what they continued to owe to the framework within which those opponents conducted their disputes. This is the skeptic attack on the whole notion of nature itself, as reported by Sextus Empiricus in his *Outlines of Pyrrhonism*. To the dogmatist claim that the senses could be trusted because nature made them coextensive with the range of sense-objects, the rejoinder was "What kind of nature, in view of the great inarbitrable disagreement among the doctrinaire thinkers about natural existence? For anyone arbitrating the very question whether nature exists would, if he were a layman, according to them be unreliable. But if he is a philosopher, he will be party to the disagreement, and himself subject to judgment, not a judge."[26]

The gist of the argument was that neither ordinary people nor dogmatist philosophers can be used to decide the issue. The former cannot, because they are just ordinary: they have no special standing in the matter. The latter are ruled out because they cannot be judges when their own opinions are what has to be judged. Pressing the point home, Sextus elsewhere questions any perception of the distinction between the natural and the unnatural.[27] "Just as the healthy are in a state which is natural for the healthy but unnatural for the sick, so too the sick are in a state which is unnatural for the healthy but natural for the sick." The argument was particularly telling against the Stoic view, in which the end or goal was often described as to "live in agreement with nature." The Stoics themselves diverged on whether this meant living in agreement with the principles that govern the cosmos as a whole or specifically in accordance with human nature, or both.[28] The skeptic attack undermined the normative, as well as the descriptive, features prominent in the use of the notion of nature from the beginning.

Where did this leave the skeptics themselves? The dogmatist views they attacked all took some position on the underlying reality of things, whether this was a matter of the qualitatively distinguishable simple bodies, or atoms and the void. The skeptics saw clearly that the arguments on either side of the dispute *undermined one another*. Their characteristic answer was to suspend judgment about reality claims altogether and to *live by the appearances*, that is, by how things seemed to the individual in the circumstances of the particular situation. This move distanced the skeptics from all those who sought positive answers to questions to do with underlying realities. The skeptics still *used* the distinction between appearance and reality, on which the dogmatists, too, depended, as the first step in any of their diverging, dogmatic reconstructions of how things really are beneath the appearances, that is, their true nature, their underlying physical constitution.

From the fourth century B.C. on, then, everyone agreed that the questions to do with elements, natures, reality, were the fundamental ones, even those who thought that those questions were strictly unanswerable. But to turn back to our fundamental problem, how far does it seem possible to account for this distinctively Greek set of preoccupations? Before we attempt some admittedly tentative suggestions on that topic, we need to say a little more on element theory in particular.

If we ask why the question of the elements of things was crucial, we should investigate what depended on being able to answer it correctly. The elements were the foundations of physical theory. If your theory of elements was sound, then the superstructure you built on its basis—the detailed explanations of particular natural phenomena—was, you hoped, well grounded. Conversely, insecurity or lack of clarity on the question of the primary elements cast doubt on the detailed explanations that presupposed them.

Here the other, mathematical sense of the term "element"— *stoicheion*—seems relevant, even though we have also to attend to the differences between mathematical elements and physical ones. The mathematical elements were the primary propositions from which, in principle, the whole of mathematics could be deduced and so demonstrated. In Aristotle's account in the *Posterior Analytics*, strict demonstration is based on three types of such primary propositions—definitions, axioms, and hypotheses. These propositions cannot be demonstrated, as he shows with an argument: If demonstration proceeds, as it must, from premises, what could be the premises from which to demonstrate the primary propositions themselves? To avoid the twin difficulties of circular proof or an infinite regress, there must be *primary* premises that are themselves indemonstrable but self-evidently true. Aristotle sets out the theory, although he views valid deduction as syllogistic in form.[29]

The best examples of such demonstrations in practice (even though they do not take syllogistic form) come from mathematics, most notably from Euclid's comprehensive treatise, called, precisely, the *Elements*. In book I, he first sets out the primary propositions he needs (in his view these are definitions, common opinions, and postulates). He then proceeds to the rigorous demonstration of most of the mathematics of his day.[30]

The mathematical elements are propositions, not physical entities. Moreover, agreement on their status as fundamental and self-evident truths was more straightforwardly attainable than was agreement on the ultimate constituents of physical objects. Yet there is an important point of contact between the two domains. In physics, as in mathematics, a concern with element theory is a concern for *secure foundations*, the firm basis on which you set out to establish the remainder of your theorems or your theories. In mathematics, the aim was strict demonstration, yielding certainty, incontrovertibility.

In other fields the goal was often the same, even though it was much more elusive.

The ambition was to produce proofs *more geometrico*, in the geometrical manner. That required not just valid deductive arguments but also primary self-evident truths, or what you could represent as such. The claims of the physicists actually to have achieved the incontrovertible were, to be sure, no sooner made than they were controverted. But they saw that to win the argument with their opponents, that was what they ideally needed. If you could claim to have secure foundations from which to demonstrate your conclusions with the certainty attained by mathematics, then there could be no further argument. Yet there always was.

The recurrent motif in our examination of three concepts—elements, nature, reality—is rivalry between those competing for intellectual prestige. To this extent, the way Greek philosophers and scientists formulated their focal problems reflected their perception of what would secure victory in that competition. It is possible to suggest a connection with, but also a break from, pre-philosophical Greek thought. Homer does not intrude in the *Iliad* or the *Odyssey*. But Hesiod, in his poems, comes forward and insists, in the first-person singular, on his own revelation, which he had from the Muses but of which he is the spokesman. He begins the *Theogony* by saying that the Muses "once taught Hesiod beautiful song, when he was shepherding his sheep under holy Helicon." He proceeds: "This word the goddesses first spoke to me, the Olympian Muses, the daughters of Zeus, who holds the aegis. . . . And they gave me a rod, cutting a branch of luxuriant laurel, marvelous to see. And they breathed into me a divine voice, so that I might celebrate things to be and things past."[31]

In a sense, as we have remarked, the contrast between appearance and reality antedates Greek philosophical speculation. But whereas Hesiod's special knowledge related to the gods and omens, the reality the philosophers claimed to reveal was the regular and therefore investigable physical properties of things. The introduction of the concept of nature thus opened up a new style of inquiry.

Like Hesiod, the early philosophers often claim their theories as their very own.[32] Neither Greek archaic poets nor early philosophers, nor, come to that, medical writers, were content to represent themselves, in the Chinese style, as the spokesmen of age-old wisdom. The Greeks did not have sage-kings to invoke as models of what it is to be wise or as the authorities for this or that item of lore or wisdom. But this in turn meant that the competition for intellectual prestige—for the position of Master of Truth—was a free-for-all. Some of the early philosophers and medical writers proceeded by insisting, against the poets or traditionalists, that the realities that counted, those worth investigating, were the natures of things. Having thereby demarcated their subject matter, they ruled out many earlier and traditional beliefs as flawed in their fundamental assumptions. The new Masters of Truth could claim to be superior because, while the opposition floundered hopelessly in the tentacles of the arbitrary, the willful, only they, the naturalists, firmly grasped what was there to be investigated.

But if investigable, also disputable and indeed disputed. The naturalists, whether philosophers or medical writers, disparaged those who depended on the supernatural. But the naturalists were far from agreeing among themselves, either on particular issues or on the nature of nature itself. True, nature provided the framework for the questions as they posed them, but they did not agree on the answers, nor did they even suggest that finding such agreement was desirable.

Clearly the circumstances of Greek intellectual life are relevant. Both in philosophy and in medicine, groups formed allegiances, and from the fourth century B.C. there were more or less well-established schools where philosophers and doctors taught. This merely provided the framework for ongoing dispute, not the occasion for the formation of a consensus. Schools, like individuals, competed with one another. What was at stake was not just reputation but also livelihood, at least insofar as those who taught, philosophers and doctors alike, depended in part on their pupils for a living.

This climate of dispute helps to explain the continuing strident claims and counterclaims to have provided new and better theories and explanations and, further, the inflation in those claims. As the criteria for success themselves became the object of reflection and analysis, nothing short of the strictest mode of demonstration, yielding incontrovertibility, was going to do. It was not enough, many maintained, for theories to be possible or plausible; they had to be shown to be true. When the game was played for those stakes, those who could claim to deliver the incontrovertibly true held the trump card.

The disagreements we have drawn attention to were not lost on the Greek theorists themselves. The Pyrrhonian skeptics, for example, as we saw, exploited those disagreements to suggest that it was wise to suspend judgment on any matter to do with underlying reality. Those who rejected the dogmatists' claims have a double importance for us. First, they confirm that some ancient Greeks, and not just modern interpreters, registered the striking extent of the disagreements on such questions. Second, even when the anti-dogmatists resisted the ambition to produce definitive theories themselves, they remained, in one essential, within the framework that Greek speculative thought had created. Their argumentative tactics were to match the dogmatists' claims against each other and so to undermine both sides in every dispute. But if that was a new way of winning the debate, their own recommendation, to live by the appearances, still used the very contrast—between appearances and reality—that goes back to archaic Greek thought.

To sum up: our first foray has suggested that certain features of Greek intellectual life, notably its fierce competitiveness, influenced the focus on certain types of questions. The ongoing disagreement on the questions in turn helped to stoke that competitiveness. Nature was the concept that enabled the early philosophers and medical writers to go one better than the poets and others who remained within the framework of traditional beliefs. Elements provided—you hoped—the secure foundation for your subsequent theorizing. To outbid the opposition, in physical theory and elsewhere, you went for the strongest claims possible, not just for the truth but for demonstrations securing incontrovertibility. Skeptics gave their own twist to the argument: they undermined all reality claims by asserting that opposing ones canceled each other out, and recommended living according to the appearances.

Causes

Our next inquiry concerns causes.[33] A first reaction to the idea of studying why the Greeks were interested in causes might well be to ask how they could have failed to be. Surely any scientist, any philosopher, anywhere, is concerned with causal explanations. Is that not the essence of any inquiry into phenomena? So how could there be a problem in understanding the Greeks' concern?

To that, the first part of the reply is to remark that there are certainly differences in degree between one society and another in the attention paid to causes, both the causes of particular phenomena and the question of what a cause is. And the second is that we can use the nature of the interest in causes and the dominant models of causation employed to throw light on the kinds of inquiry considered important and on the characteristics of the science conducted on their basis.

There are some important differences too between ancient Chinese and Greek investigations, not that there is a polar contrast between a Chinese interest in correlative thinking and a Greek interest in causation. On both sides that contrast would be flawed. The Greeks had a developed interest in correlations. Tables of Opposites, like the one attributed to the Pythagoreans by Aristotle, are only one example.[34] The Chinese investigated causes in a variety of contexts—for example, in connection with medical diagnosis and with human responsibility for the success or failure of policies. Except for some elliptical remarks in the Mohist Canons, however, the extant Chinese texts from the Warring States and Han periods do not contain explicit analyses of causation.

The Greek preoccupation with causal accounts, and with the nature of causation itself, is, by contrast, intense. Why should this have been so? What were the main models of causal explanations that Greeks adopted?

The first extensive evidence comes from the attempts to distinguish different types of cause in some of the Hippocratic treatises dating from the fifth century B.C. The need to distinguish between what was responsible for disease and what were concomitant or coincidental factors was remarked in such works as On Regimen. The author of On Ancient Medicine specifies that the causes of each disease must be those things whose presence necessarily brings about the disease and without which the disease ceases.[35]

By the fourth century B.C. Plato made an analysis of different types of causal factors a central argument when advocating the importance of what he called Forms. In his cosmological dialogue, the Timaeus, he insists that a proper explanation of the world must be in terms of the good that it manifests—the work of the figure he calls the Demiurge or Craftsman, who imposes order on disorder. The physical factors that bring about heating and cooling, solidifying and dissolving, and so on, are merely auxiliary causes (sunaitia), although other people, he remarks, consider them proper causes (aitia).[36]

Similarly, in the more down-to-earth context of explaining why Socrates, condemned to death by the Athenians, chose to remain in prison awaiting execution rather than to accept his friends' offer to help him escape, Plato makes Socrates insist in the Phaedo that this is because of his sense of what was right.[37] The material constituents of Socrates' body, the physical arrangement of his bones and sinews, cannot account for his sitting there; they are merely the necessary conditions without which the true cause could not operate. Plato undertakes this excursus into what makes a cause a cause partly to undermine the view of those who privileged physical interactions in their explanations of objects or events. Because they disagreed on what counted as a cause, discussion could not concentrate on assigning particular causes in particular cases. The fundamental issue was what types of account could be valid.

The polemical function of what might otherwise seem a neutral analysis of causation is even clearer in Aristotle. He set out to review everything his predecessors had had to say on the subject and claimed that his own view went beyond, and superseded, all of theirs. His own theory of the four causes, material, formal, final, and efficient, owed a good deal more to his predecessors than he let on, in particular with regard to the final and efficient causes, both evident enough in Plato. Aristotle was the first to set out a formal taxonomy of causes, specifying the questions to be investigated. The effect of this was not just to systematize earlier reflections; by clarifying the questions, that is, by redefining them, he took a large step toward ensuring that his public would see his own answers as superior.

As with Aristotle's theory of the material elements, it is not as if his view subsequently won out over all its rivals, let alone achieved the status of an orthodoxy. Both the main positive Hellenistic philosophical schools contributed to the ongoing debate.

First the Stoics put forward a sophisticated alternative to Aristotle's analysis, for which Chrysippus was largely responsible.[38] We cannot go into the details here, but he was concerned, among other things, with the distinction between factors that can be removed without removing the effect and those that cannot. He does not have the terminology of necessary and sufficient conditions but explores some of the distinctions that those terms can be used to express. The Stoics were not fatalists, but they were determinists, maintaining that every event is fully determined by its causes and that the universe manifests an unbroken nexus of cause and effect.

The Epicureans, by contrast, insisted that there are breaks in that nexus. To win a place for free will, they postulated an uncaused event—the swerve of atoms. They diverged fundamentally from Platonists, Aristotelians, and Stoics alike in denying teleology, the explanation of things in terms of the good. Whereas they thus rejected final causes, they retained other types of causal explanation. Happiness and peace of mind depended on understanding causes, as we have seen. You had to know that the ultimate reality consisted of nothing but atoms and the void. And you had to appreciate that even the most obscure, strange, or frightening phenomena had natural causes. Indeed, the principle of plural explanation dictated that no causal explanation that seemed *possible* should be rejected. Long after the astronomers had arrived at what they accepted, with good grounds, as the correct explanation for lunar and solar eclipses, Epicurus and, after him, Lucretius continued to contemplate all sorts of implausible accounts—for example, that the heavenly bodies were temporarily extinguished.[39] Yet this anti-dogmatism still seemed essentially dogmatic to the Pyrrhonian skeptics, for whom any account of hidden causes was to be avoided.

The parties to the disputes on the nature of causes appreciated that the issue was not just the analysis of cause but the very agenda of science and philosophy. What types of explanations are feasible, they asked, not just in relation to the phenomena of astronomy, physics, or medicine, but with regard to the universe as a whole? A key issue was teleology: Is the universe controlled by a divine, benevolent, craftsmanlike force or not? Both those who argued for and those who argued against such a position saw that the prior question was the applicability of final causes to cosmology.

These arguments about causation provide yet another example of the Greek propensity for debate. The next question is, How far was this preoccupation related to particular features of Greek intellectual or cultural life? What might it owe to the intense involvement of Greek citizens, in the classical period especially, in the law? As we explained in Chapter 3, they sometimes acted as legislators responsible for constitutional decisions, as litigants in civil or criminal cases, and as dicasts serving as both judge and jury when cases were tried.

The original connotations of the main Greek terms for causes, *aition*, *aitia*, and their cognates, link them firmly with the legal context, not just with the domain of human behavior in general. *Aition* denotes what is responsible for something. *Aitios*, in the masculine, is used of the guilty party. *Aitia* means "blame or guilt, its apportionment, or an accusation imputing blame."

An interest in responsibility and in apportioning blame antedates the development of the elaborate legal institutions, such as the Dicasteries, of the classical Greek city-states. This can be seen from such examples as the famous passage in the *Iliad* where Agamemnon, explaining why he took Briseis from Achilles, says: "I am not to blame for [literally, "I am not the cause of"] this act; rather, Zeus and my portion and the Erinys who walks in darkness are." Although Agamemnon says that Zeus was at work, he does not evade his own obligation to pay compensation to Achilles. To that extent and from that point of view he accepts his own responsibility.[40]

Further light is shed on the archaic or traditional notions and on their transformation by considering the use of *aitios* and its cognates in the Hippocratic treatise *On the Sacred Disease*, already cited in connection with the early concept of nature (*phusis*). Although *phusis* marks out the Hippocratic writer's position from that of the purifiers he attacks, the terminology of responsibility and causation is used *both* in the views he rejects *and* in the account he advocates to replace them. The purifiers, he says, explained different types of the "sacred disease" (epilepsy) by assigning responsibility to different gods. Thus, "if the patient imitates a goat, or roars, or has convulsions on the right side, they say that the Mother of the Gods is to blame (*aitien*). If he utters a sharper and louder cry, they liken him to a horse and say that Poseidon is responsible (*aition*)."[41] But whereas

the purifiers concentrate on determining which divinity is personally responsible, the Hippocratic writer himself insists that the disease
is natural and has a natural cause. He uses *aitios* and its cognates, along
with the terms *phusis* and *prophasis* (cause), for his view as well as
theirs.

With this view we might compare what happened to the
dichotomy of appearance and reality when the naturalists took it
over. That things may not be what they seem had been appreciated
long before. In the naturalists' hands, the underlying reality became
a matter of the essential natures of things. Similarly, the question of
responsibility was no longer a matter of holding persons, human or
divine, responsible for diseases but of assigning the true, regular
natural causes to them.

The naturalists thus appropriated and depersonalized the topic of
responsibility. Causal explanations still meant identifying what was
responsible for the effects to be explained, but they placed it outside
the domain of personal agency and in that of the intrinsic properties of things. In this program, the natures of things are the goals of
inquiry, yet the idea of responsibility is still present. In a sense, nature
itself is responsible.

The heightened interest in causation found in philosophy and
science points, then, to the general influence of Greek legal experience on Greek culture. That interest sprang partly from the concern
with questions of responsibility in the legal domain. We must qualify
the idea, to be sure, in three respects. First, the influence may well
have been two-way, not just from the legal experience to the philosophers' theoretical reflections but from the latter to the sophistication
with which causal issues came to be discussed in the actual practice
of forensic disputes. Second, other fields of experience had important contributions to make; we will be coming back later to the roles
that models drawn from the fields of art and technology and from
living creatures play in cosmology. Third, some Greeks reacted in a
highly critical way to the models of advocacy associated with the law

courts and with rhetoric more generally. Both Plato and Aristotle con-trasted proper philosophy, intent on the truth, with techniques for persuading an audience in the legal and political spheres.[42] What seemed plausible to a group of dicasts or to the demos in Assembly was not necessarily what was true or what was good. For phi-losophy, what counted was not the votes of the majority, but what was really the case.

If the law could provide positive models for intellectual inquiry, it could also be the source of negative ones—of the dangers of merely persuasive arguments and the deceptions of rhetoric in general. But the ambivalence goes deeper still. The ideal that both Plato and Aristotle presented was one of rigorous demonstration, apodeixis, rather than mere persuasion. In Aristotle's case, the strictest mode of demonstration proceeded by valid deduction from self-evident primary premises, its axiomatic basis. Demonstration was still a matter of exhibiting the causes of things, incorporated, in Aristotle's view, in the middle terms of syllogistic proofs. Insofar as those aitiai still carried the connotations of responsibility, they had that much in common with the subject matter of legal debates. However, they differed from that subject matter in that it was no longer guilt or innocence that was at stake.

The term used by Plato and Aristotle for demonstration is yet another example of the appropriation and transformation of a term that was central to the opposition's vocabulary. From the earliest extant examples of Greek forensic oratory, from Antiphon and Lysias in the fifth century B.C. to Demosthenes and Isocrates in the fourth, the orators spoke freely of what they believed they had established in terms of "proofs," using the same terms—apodeixis, apodeiknumi, and cognates—that the philosophers later deployed when they set out to define the strictest type of demonstration.[43] Of course, what the orators showed and the way they showed it were very different. Their aim was establishing the facts of the matter and the guilt or inno-cence of the parties concerned. Their proofs were not strict deductions from primary indemonstrables but what could be represented as having been established beyond reasonable doubt. Although the philosophers set their sights on the positive ideal, in redefining "demonstration" they used as negative models the very modes of argument they sought to downgrade.

The contrasts with the Chinese experience, on both the positive and the negative side, are fundamental. Chinese philosophers and scientists never attempted to cultivate demonstration in the axiomatic-deductive mode, nor did Chinese legal experience provide rich sources of alternative models of forensic argument.

The influence, both positive and negative, of Greek legal and political experience thus had a variety of important manifestations. A final example might be the elaborate analogy that the astronomer and harmonic theorist Ptolemy set up in his short epistemological treatise On the Criterion.[44] Throughout his scientific work his methodology aims to combine reason and experience, the criteria that had been the focus of epistemological debate from the Presocratic period on. In On the Criterion one of the ways he illustrates their relation is through the analogy of a trial. He compares the basic perceptibles to those who are judged, and reason to the judges. It is as if he found it natural to construe any inquiry into the phenomena as an attempt to determine what was responsible by examining the testimony of rival witnesses, the whole having a form and outcome like those of a due process of law.

Mathematical and Physical Explanations

The issue that this section addresses is suggested by two points that have already been made. We have seen that Aristotle lays down among the conditions for strict demonstrations that they should proceed by way of valid deductive argument from indemonstrable starting points that are themselves self-evidently true. In practice, however, we do not find such demonstrations set out in his own investigations of animals, for instance, where it is far from clear what would count as axioms on a par with the equality axiom in mathematics. That states that if equals are taken from equals, equals remain—a principle that is indeed self-evident and could not be proved without circularity. Some have suggested that the Aristotelian principle that nature does nothing in vain counts as such, in that he believed that without an adherence to some such principle—to the validity of final causes, in other words—no investigation could make any progress. Yet Aristotle certainly knew that that principle was quite unlike the equality axiom in this respect, that it was highly disputed, and indeed, he devotes many pages to trying to establish it—which is not something that anyone would need to do with the equality axiom, for sure.

Greek mathematics, which may or may not have been influenced by Aristotle, does set out axiomatic-deductive demonstrations of mathematical theorems, with Euclid's own *Elements* as the most comprehensive and systematic example.[45] These demonstrations are not syllogistic in form, but they are deductive, and they proceed from explicit starting points (the elements themselves) that are indemonstrable and self-evidently true. To be sure, some questioned the status of some of Euclid's indemonstrables. That happened most notably in the case of the parallel postulate, which states that nonparallel straight lines meet at a point—the fundamental spatial assumption on which what we call "Euclidean" geometry rests. Some ancient commentators said that that should be a theorem to be proved, although all attempts to do so turned out to be circular.[46] Euclid's own position was not just that it had to be postulated but also that it is true.

The juxtaposition of Aristotle's zoology and Euclid's *Elements* might suggest a radical distinction in the styles of explanation proper to the natural and the exact sciences. We might suppose that the Greeks would have adopted a similar contrast, taking exactness to be attainable in mathematics but not in the study of nature. The latter, in tackling the problems set by the empirical phenomena encountered in zoology, botany, meteorology, and so on, deals with what is (as Aristotle put it) true "for the most part," as well as with what is true always. Mathematics can be exact only because it deals with abstractions—ideal situations in which any physical aspects of the subject under consideration have been discounted.

But that was not at all the way the Greeks as a whole saw things. Understanding why throws important light on their aims and ambitions in the investigations they undertook. Two of the most highly developed branches of inquiry—medicine and astronomy—show the complexity of their responses.

First, however, we need to mention that some Greek writers did maintain that the price paid for the exactness of the exact sciences was their abstraction from the physical aspects of the situations they dealt with. Archimedes, often held up as emblematic of Greek science as a whole, is in many respects, and in the one that concerns us here, exceptional. Most of his extant work deals with problems in pure mathematics (to use the modern term), having to do with the surfaces and volumes of spheres and cylinders in *On the Sphere and Cylinder*, for instance, and the areas of segments of a parabola in *On the Quadrature of the Parabola*, and the properties of spirals in *Spirals*. But then there are his important forays into what we would call statics and hydrostatics, in *On the Equilibrium of Planes* and *On Floating Bodies*.

Two features of these works are crucial for assessing how he achieves his results. On the one hand, in both statics and hydrostatics he incorporates certain physical terms in the statements of the axioms or postulates. Thus the first postulate in *On the Equilibrium of Planes* specifies that "equal weights at equal distances are in equilibrium."[47] On the other hand, in both he idealizes the phenomena. The lever imagined in *On the Equilibrium of Planes* is not a physical—for example, metal—bar; Archimedes discounts its weight and flexion and the friction that would accompany its movement about a fulcrum. In *On Floating Bodies* he stipulates in the postulate with which the book opens that the fluid is perfectly homogeneous and totally inelastic. Thus, both studies proceed, after the initial physical references, in a purely mathematical fashion. Even if they are (loosely speaking) bids to mathematize physics, we must be clear that the elements of physics involved are minimal, and that this is deliberate, for the sake of the mathematical analysis. But if, in Archimedean statics and hydrostatics, the physical aspects of the problems figure almost exclusively in the postulates, other investigators did attempt to express physical explanations mathematically.

The two subject areas we have chosen for special investigation are medicine and astronomy, and both illustrate the tensions that arose from, on the one hand, the desire for proofs in the geometrical manner and, on the other, the difficulty of attaining them while paying due attention to the physics of the problems to be solved. We may concentrate, for our purposes, on the extensive and articulate writings of Galen, on the one hand, and of Ptolemy, on the other.

Attempts to model medicine on an exact science antedate Galen. The possibility was already a subject of controversy in the Hippocratic Corpus, where the author of On Ancient Medicine protests that medical practice is inevitably inexact.[48] Once Praxagoras of Cos became the first Greek to make the the pulse important in diagnosis (around 300 B.C.), the development of pulse theory in such writers as Herophilus owed much to the models provided by music theory. Galen reports that "just as musicians establish rhythms according to defined sequences of time units, comparing the arsis (upbeat) and the thesis (downbeat) with one another, so, too, Herophilus supposes that the dilation of the artery is analogous to the upbeat, whereas the contraction is analogous to the downbeat." The ambition evidently was to give pulse theory a mathematical analysis analogous to the successful analysis of the musical concords in harmonic theory.[49]

Galen himself went much further in stating and attempting to implement the ideal of proofs in the geometrical style in medicine.

His views are all the more impressive because, first, he was a logician of considerable caliber—although his magnum opus in this field, *On Demonstration*, in fifteen books, is no longer extant. As an experienced physician, second, he points out in his pharmacological writings such problems as the differences in the properties of drugs prepared in different ways or used in different combinations and the difficulty of predicting their effects on individual patients with different constitutions. In contexts such as these he repeatedly emphasizes the lack of exactness.

That did not stop him from demanding that, ideally, arguments in medicine and in physiology should be as certain and as exact as those in geometry. The general problem this poses is clear.[50] What would count as the self-evident primary indemonstrable premises in those fields? One of Galen's examples was the principle that "opposites are cures for opposites." The problem here was what is to count as an opposite. Even such apparently simple qualities as hot and cold, wet and dry, were, as Aristotle had long ago pointed out, hard to define and controversial. Galen himself distinguished various degrees or grades of heat, cold, and so on, which aimed, to be sure, at quantitative differentiations but reduced, in practice, to impressionistic qualitative ones. Yet if the opposites are merely stipulated to be what produces a cure, that makes the principle true but vacuous.

Nor does Galen do much better in his search for geometrical-style proofs of other theories such as that the brain is the source of the nervous system and the heart of the arterial one. True, he can demonstrate, in the sense of exhibit, the effects of certain interventions—for example, what happens when the nerves or the arteries are ligatured. But while, we may say, he has good evidence and arguments for his conclusions, that does not justify any claim that they have been demonstrated from self-evident axioms. Here is a clear instance where, in possession of good *empirical* grounds for his theses, Galen nevertheless hankers after the mode of certainty and exactness that only the geometrical model could deliver.

The second field we chose to test Greek ambitions was astronomy, the study of the heavens, where mathematics always had a far greater role to play than in medicine.[51] From the fourth century B.C. on, Greek astronomers attempted one geometrical model after another to account for the movements of the sun, moon, and planets. This was not just with the idea of describing the movements and determining concrete parameters to predict them. The aim was also to explain the movements by asserting that they were caused by the celestial spheres of which the models gave the geometrical properties. Not all Greek astronomical writing proceeds to the point where it explicitly translates the geometry into a physical theory.[52] Sometimes, however, it does. In the first clear example of this, in Aristotle, the physics in question is exceptional, for he thought of the heavens as composed of a fifth element, *aither*, quite unlike the sublunary four (he needed it to account for circular movement). His celestial spheres were transparent—indeed, invisible— and needed no push to keep them going forever: they were indeed divine.

It is also true that some extant treatises of Greek astronomers deal purely with the mathematics of the problems posed. Autolycus, in *On the Moving Sphere*, for example, studies interacting great circles on a rotating sphere as a purely geometrical issue. The one work of Aristarchus of Samos that has survived, *On the Sizes and Distances of the Sun and Moon*, does not, despite its title, give concrete results for those magnitudes but concentrates rather on exploring the geometrical conditions for solving the problem.

Whether Aristarchus's heliocentric hypothesis is a further example of a purely mathematical study is controversial. We know on impeccable authority—that of Archimedes no less—that Aristarchus proposed a number of theses which included the suggestion that the sun is at the center of the planetary system and that the earth revolves around it like one of the planets.[53] Some ancient commentators, such as Plutarch, interpreted this as put forward merely for the sake of argument, although Plutarch knew of another ancient astronomer, Seleucus of Seleucia, who adopted that hypothesis as the basis for a true physical account. Yet Plutarch's view of Aristarchus's own position is open to question: it hardly explains the adverse reaction of some contemporaries. The Stoic philosopher Cleanthes is reported to have said that Aristarchus ought to be indicted for daring to shift the earth from the central position in the universe—not that any indictment took place.[54]

Although there is some doubt about the exact status of the original heliocentric suggestion, there is none about the views and objectives of the most authoritative and best documented ancient Greek astronomer, Ptolemy. Three points are fundamental.

First, Ptolemy himself provides a physical and dynamic equivalent to his mathematical models in his work *On the Planetary Hypotheses*. The models, based on epicycles and eccentrics, are worked out in the *Syntaxis*, which says nothing about what they correspond to or are made of. But in *On the Planetary Hypotheses*, Ptolemy interprets them physically in terms not of entire spheres but of tambourine-shaped strips cut from them. He offers a vitalist account of their motions: they move because they are alive.[55]

Second, in the *Syntaxis* itself, he firmly bases his whole account on certain physical assumptions, chief among them the thesis that the earth is at rest in the center of the universe.[56] He does not take this principle, important as it is, for granted. On the contrary, he mounts a battery of arguments to recommend it. Most are, moreover, physical arguments—for instance, the appeal to the Aristotelian doctrine that heavy objects naturally move toward, and come to rest at, the center of the universe, and the absence of effects that he says one would expect if the earth was rotating once every twenty-four hours. Clouds, for instance, could never move eastward, because they would always be anticipated by the motion of the earth.

The third and most important point relates to Ptolemy's remarks on the status of his astronomical theorizing. The opening chapter of the *Syntaxis* could not be more explicit.[57] He contrasts "mathematics"—including the astronomy that is to follow—with both "physics" (here the study of changing material objects) and "theology" (the study of the gods, which some saw as the most important part of philosophy). Both the latter are conjectural; the trouble with physics, on this view, is that it deals with what is changing and unstable. But mathematics, by contrast, holds out the promise of firm and unshakable knowledge, for its proofs come about by the incontrovertible methods of arithmetic and geometry.

Thus, Ptolemy marks off the study of the heavenly bodies from that of terrestrial phenomena. In the opening of his astrological treatise, the *Tetrabiblos*, he further distinguishes the two main modes of the study of the heavens, mathematical astronomy and horoscopic astrology. Predicting the movements of the heavenly bodies can and should be a matter of demonstration, because it deals with what is unchanging, but the study of what they portend for events on earth is, once again, a matter of conjecture.[58]

In practice, much of the *Syntaxis* is, we should say, steeped in inexactness.[59] Ptolemy uses for important parameters values that he knows to be only approximate and frequently rounds figures in his calculations. He does not allow these compromises to count against his claims for the subject as a whole. Its axioms are not the self-evident truths of Euclid's *Elements* but physical principles that have to be argued for. Yet it is a demonstrative study, deploying incontrovertible proofs. As in the very different circumstances of Galen's biology, the ideal of mathematics enthralled Ptolemy, to the point where he treats his astronomy *as* mathematics rather than physics, despite the physical assumptions on which it must rest.

We began this section by remarking that the Greeks could contrast the exact and the natural sciences by means of the distinction between mathematics and physics. Yet we have found that different theorists used that distinction in markedly different ways. Not everyone recognized that the price to be paid for the adoption of the Euclidean mode of proof was abstraction from the physical aspects of the problems—on the grounds that self-evident axioms are possible only in pure mathematics. Some prominent Greek theorists did not accept that limitation. Both Galen and Ptolemy, in their different ways, sought to extend the model into the areas of physics they were interested in. They wanted incontrovertibility far beyond the domain of the purely mathematical.

From one point of view, this might seem merely to reflect a desire for the highest possible standards of rigor. But from another, taking into account the characteristics of Greek intellectual exchanges, there is more to it than that. What drove this ambition was competition in dialectical debate, the recognition, made explicit in Galen especially, that to win the argument against your rivals you needed to claim not just that truth was on your side but also certainty.

In China, too, the study of the heavens, especially, depended on the rigorous quantitative analysis of carefully collected and tested empirical data. But although classical Chinese astronomers continually sought greater accuracy in establishing eclipse cycles and other calendrical constants, for example, they did not attempt to prove a view of the physical dispositions of the heavenly bodies, let alone show that things must be so and could not be otherwise. They did not seek to set aside the physical aspects of the phenomena in order to obviate their inexactnesses. They did not, therefore, encounter the difficulty that some Greek investigators did when they tried to find axioms that were certain and not empty. The Chinese did not feel a need for incontrovertibility, the driving force in such Greek investigations.

This section has mostly taken up questions to do with specific domains of inquiry. It remains for us to investigate Greek attempts to give comprehensive accounts of the world as a whole, in other words their cosmologies.

Cosmos: Assumptions and Debates

Of the topics we have chosen for study in this book, the cosmos is the most promising candidate for the claim that at least the *questions* in ancient China and in ancient Greece were broadly similar. We might expect this to be the case with respect to the sky and the earth and thus to the universe as a whole—to cosmography and cosmology.

We should remark several important, indeed fundamental, simi-larities at the outset. First, both Chinese and Greek ideas on these topics are deeply value-laden, although the values greatly differ. In neither case is cosmology divorced from the domain of the moral and political. Second, ideas about the macrocosm mirrored and were mirrored in ideas about the two microcosms of the body and the state. This was sometimes a matter of direct comparison or analogy between these three domains and sometimes a conviction that they constituted a seamless whole. Either way, there is a pervasive inter-action of the ideas deployed in the understanding of all three. Third, concepts of the cosmos drew largely on notions of harmony and good order or governance, although the forms that those ideas took were far from identical in the two civilizations. Superficial similari-ties in the terminology should not lead us to ignore differences not just in the answers given but also in the understanding of the ques-tions themselves.

Although the earliest extant Greek literature contains material from which scholars can reconstruct a cosmographic picture of a sort, it presents no cosmology in the sense of a vision of the uni-verse as a strongly unified whole. Homer refers to a domed and apparently solid heaven over an earth surrounded by Ocean; below the earth was Hades, the abode of the dead, and Tartaros was below that. Hesiod corrected Homer's picture—for instance, on the ques-tion of just how far below the earth Tartaros was: a bronze anvil falling from heaven would take nine days to reach earth and another nine to arrive at Tartaros.[60] But this, too, is more like cosmography than cosmology.

The Presocratic philosophers were the first to embark on theorizing about the universe as a single ordered whole. As explained previously (p. 146), the recasting of their ideas in later reports, the only ones available to scholars today, often makes their original form problematic. Once again, however, these early philosophers provide the background to later theorizing.

The central term round which much later Greek cosmological speculation revolved, *kosmos* (world-order), appeared in Heraclitus around 500 B.C. "This *kosmos*," he says, "the same for all, none of the gods nor of humans made: but it was always and is and will be an ever-living fire, which is being kindled in measures and extinguished in measures."[61]

The verb *kosmein* and the noun *kosmos* already occur in Homer in the senses of "order, arrange, arrangement" and "adorn, adornment." Heraclitus speaks of the world as a whole as ordered: *kosmos*, here, does not just mean "world," as it often did in later writers, but, precisely, world-*order*. Two other motifs prominent in later cosmological speculation also appear here: the representation of the world as alive (an ever-living fire, indeed) and the denial that it was created. The first two Milesians, Thales and Anaximander, may have been more concerned to account for the origins of things—with cosmogony, in other words—than with element theory. If that is correct, then Heraclitus's fragment would gain particular point, in that he implicitly denies that the world-order came to be.

From the Presocratics down to the end of Greco-Roman antiquity we find one cosmological account after another. Most writers agreed that some account of the world *as a whole* was necessary, and some argued that perception gave access to no more than a realm of appearance. Parmenides, for example, strongly contrasted what he called the Way of Truth with the Way of Seeming, the former based on reason and argument alone. Yet he had the goddess responsible for both accounts set out, in the latter, a cosmology in order—as she puts it to the youth to whom the poem is addressed—that "no opinion of mortals will outstrip you."[62] Even while he undermined the cosmological project as no more than a likely story, Parmenides thought it worthwhile to propose a better cosmological account than any of his predecessors. Some likely stories, he implied, are more likely than others.

As with element theory, in cosmology the debate between rival theories was intense. What was at stake was the best *comprehensive* account of the universe. The ambition to provide such an account united philosophers even while their competitiveness ensured that the views they put forward differed widely. Is the universe eternal or created? Is there just one world, several, or an indefinite number, and if more than one, are these worlds separated in space or in time, or in both? Again, are matter, space, and time all infinitely divisible continua, or are they constituted by indivisible, atomic quanta?

Some of these positions cluster with others.[63] Those who postulated physical atoms tended to be indivisibilists on questions to do with space and time as well, and often maintained a plurality of worlds. Even so, the combinations and permutations of such positions were very great in number. Some theorists offered their views merely as probable accounts, but the majority wanted more and claimed theirs to be correct. Even among those who thought probability was the most that could be attained in cosmology, some did not hesitate to demand certainty elsewhere, as Plato did with dialectic and the study of the intelligible Forms.

Although some investigators attempted to bring empirical considerations to bear on particular issues—as when Hero of Alexandria in the first century A.D. set out certain tests to show that a vacuum can be produced artificially—in the main the weapon used on either side of the cosmological disputes was abstract argument.[64] One typical thought experiment, used to suggest that the universe is spatially infinite, may go back to the pre-Aristotelian Pythagoreans (it was certainly employed later, in various versions, by both Stoics and Epicureans). Archytas, in the fourth century B.C., appears to have asked: "If I were at the extremity—say, at the heaven of the fixed stars—could I stretch out my hand or staff, or could I not?" It would be absurd to think he could not do so.[65] But that means that there will be either body or place outside the supposed extremity—and the argument could be repeated for any further extreme point that might be suggested.

Three main models or analogies permeate Greek cosmological thinking: the vitalist, the technological, and the political.[66] We now need to study some of the varieties of each and the underlying issues that were at stake. The first represents the world as, or as like, a living being; the second, as, or as like, an artifact; and the third, as, or as like, a state. In each case it is important to distinguish, on the one hand, a comparison or an analogy—between the universe and one or other of these three types of entity—and, on the other, an assertion that the universe is just such an entity. A comparison or analogy may implicitly distinguish the universe from what it is compared to; an assertion treats the universe as an instance of a kind of entity.

The conceptual distinction is clear, but in practice the issue may be blurred. When theorists say that the universe is like a living being, they may not mean to deny that it is a living being; they may mean only that it is no ordinary one. Or they may intend to contrast the universe as the quite exceptional living being it is with common animals or plants. It resembles them in being animate, although the way it is animate differs from the ways animals or plants are alive. Some Greek cosmologists appear to switch between identity statements and comparisons, although both types of account have in common that they apply vitalist, technological, or political ideas, as the case may be, to representing the universe as an ordered whole.

All three types of account figure already in the Presocratic period, but in the fourth century B.C. they became the focus of cosmological controversy. On the one hand, Plato, Aristotle, and the Stoics used all three types in varying ways to advocate a teleological view of the universe. On the other, the anti-teleological Epicureans constructed their cosmologies using some of the same basic ideas.

At the end of the *Timaeus*, Plato describes the cosmos as a "visible living being, encompassing the visible living beings, a perceptible god, the image of the intelligible one"; and elsewhere his account draws extensively on both technological and political terminology. He describes the divine figure responsible for imposing order (*taxis*) on disorder as the Craftsman and repeatedly captures his activities in the vocabulary of the arts and crafts: for example, the Craftsman acts as a carpenter, as a wax modeler, and even as a baker. But he is also supreme ruler, in control of the universe, said in the *Philebus* to be king.[67]

According to Aristotle, there is a craftsmanlike force at work in the universe, not a transcendental one like Plato's but rather one immanent in nature. He, too, is emphatic, in the *Metaphysics*, that there is a supreme principle of order in the world, the Unmoved Mover. Although he criticizes those who assume that the material elements of things are alive, he thinks of the heavenly bodies as living and is convinced that the earth is subject to cycles of growth and decay.[68]

Finally, our secondary sources report that the Stoic Zeno maintained all three ideas, that the cosmos is a living creature, that it is governed by providence and divine law, and that a craftsmanlike, purposeful activity is at work throughout nature.[69]

The anti-teleologists, for their part, avoided the features of these models that implied design, purpose, providentiality. Yet the adaptability of vitalist and technological terminology was such that it could be used without the notions that dominated in the teleologists' applications. This can be seen already in the Presocratic period where the earliest atomists, Leucippus and Democritus, used terms taken over from embryology to represent worlds being formed. They spoke of the cosmos being enclosed in something they compared to a membrane, or as arising from a seed mixture, yet they did so without thinking of these *kosmoi* as designed.[70]

The Epicureans spoke of worlds as *kosmoi*, of their being made— though not by a Craftsman—and of their growing, except that now this happens because of the atoms in their constant movements and interactions. In the *Letter to Pythocles*, for instance, Epicurus speaks of seeds rushing together to form a world, and when he adds that they receive "waterings" from appropriate sources, he evidently has in mind the growth of plants. Yet these seeds—like those that account for the generation of animals—are not themselves agents with intentions, nor are they evidence of any intentionality at work in the scheme of things.[71]

The great variety of Greek cosmological accounts is to be expected, in view of the systemic competitiveness of Greek philosophy and science. As with the dispute over what constitutes a cause, we can see that not just the answers but the conceptions of the question differed. For the teleologists the term *kosmos* itself carried with it associations of good order and design that the anti-teleologists specifically rejected. What both parties had in common was the ambition to produce a comprehensive account of all there is. The teleologists saw the problem as one of investigating not just the regularities of cosmic processes but the good they manifest. The Epicureans denied the latter and concluded from the principle that both atoms and the void are infinite, that "the number of worlds— both like this one and unlike it—is also infinite."[72]

The use of political models in cosmology offers a promising opening for the investigation of the interactions between Greek thought and Greek society. For the teleologists the world was in a sense a monarchy, although the notion of the cosmic ruler and his rule differed as between Plato's Craftsman, Aristotle's Unmoved

Mover, and the Stoic divine reason. At the opposite end of the political spectrum (as it were), Heraclitus saw the world as anarchic when
he claimed that "justice is strife" and that "war is father of all and
king of all."[73] That contradicted Anaximander's view that certain
cosmic forces "pay the penalty and recompense to one another for
their injustice, according to the assessment of time." Anaximander
suggested, rather, a balanced, just relationship between equals, a
limited democracy or an oligarchy—a conception also found in
Empedocles, for example.[74] Even the Epicureans used images with
political associations when they pictured the gods (who, they
insisted, have no effect on this or any other world) as enjoying a bliss
that Lucretius, for instance, describes as *peace*.[75]

These diverse cosmological images reflected the variety of Greek
political experience. Indeed, the disagreements about the cosmic dispensation mirrored the disputes both in theoretical political philosophy and in practical politics over the ideal constitution. From at
least Herodotus onward, the competing claims of democracy, oligarchy, and monarchy were at the center of those disputes. That continued to be the case in political philosophizing long after the
city-states had lost much of their autonomy.

In this regard, a gulf separates the two ancient societies of Greece
and China. There was plenty of turbulence in Chinese political life,
both before and after the Ch'in unification. But the agreed ideal
remained the benevolent rule of a wise prince as guarantor of
harmony between heaven and earth. Although not many princes
turned out to be wise, the Chinese did not experiment, in theory or
in practice, with other political constitutions.

In Greece, by contrast, all was rivalry and dispute. The variety
of political constitutions did not inevitably lead to the use of each
in cosmology. But Greek competitiveness was at work in both
domains, for in both politics and cosmology it positively favored
trying out radical new proposals, both in thought and theory and in
action.

A point of similarity between China and Greece, already noted, was the conception of cosmic order as a matter of good government and harmony. Whereas, at one level, Greek cosmologists exploited the considerable variety of models that political experience exemplified, at a deeper level, all of them—democratic, oligarchic, monarchic, even anarchic—shared the notion that order is a matter of rule. That idea was present whether the political image was that of the rule of one or of many. It can be found even in the version of cosmic anarchy proposed by Heraclitus, when he put it that war is king. Despite their diversity in other respects, these cosmologies were all, in one way or another, based on the contrast between what rules and what is ruled. This might suggest that the real point of similarity between Chinese and Greek cosmology was rooted in that contrast, so that both reflected a deeply hierarchical view of reality.

Such a thesis has its attractions. But it stimulates further analysis of what notions of hierarchies were in play in China and in Greece. The emphasis in China was regularly on the interdependence of opposed, complementary factors in political, social, and familial relations and in the manifold other applications of the fundamental yin and yang principles. The dynamic harmony that the Chinese saw as the proper relation between heaven and earth was a matter of their complementarity.

Some Greek views of hierarchy—and some of the most influential ones—postulated a different ideal, not the interdependence of the higher and the lower but of the independence of the former from the latter. The Greeks realized that male and female need one another for the purposes of reproduction. But when Aristotle suggests that those species of animals in which male and female are distinguished are superior to those that generate asexually, he explains that the separation of the sexes liberates the male to be able to perform his higher functions.[76] In Greek political life, the roles of men and women were markedly different, in that even though women had certain civil rights, the political functions of citizens were the purview of men alone.[77] Yet Aristotle carried this social and political contrast over into his zoology.

The free members of Greek city-states—citizens and resident aliens—depended heavily on the labor of slaves in agriculture, technology, and every area of economic life to a far higher degree than was the case in ancient China, even allowing for the differences in unfree status in the two (p. 21). Yet Greeks liked to think of the free as independent and, in presenting that image, usually ignored the real-life interdependence of free and slave.

Where higher political activities were concerned, the Chinese regularly stressed the mutual dependence of rulers and ministers and the ways the prince should heed advice. In the ancient Greek city-states there were no ministers, no stable positions of influence that their holders could use to sway the policies of those in power. In the democracies the next vote of the Assembly could overrule what the last one had decided. Even when the Greco-Roman world eventually had emperors, surrounded by courtiers and served by a bureaucracy, these were Romans, not Greeks, and even some of those who saw the advantages of imperial rule harked back to republican virtues.[78]

Even though ruling was such a dominant image of orderly arrangement in both ancient societies, the differences between Chinese and Greek hierarchies are clear enough. Whereas yin and yang are essentially interdependent and defined in terms of each other, Greeks more frequently stressed the opposition between members of pairs even while admitting their interrelation. Both cultures prized musical harmony, with its due balance between the high and low in pitch. Whereas the Chinese saw cosmic harmony as a matter of the due relation between heaven and earth, the Greeks tried out a variety of ideas according to which the heavenly spheres were themselves harmonious.[79] The harmonies came from the movements of the spheres. Whereas the Chinese made the emperor ensure the harmony between heaven and earth, the Greeks viewed the harmony of the spheres as independent of any human activity or responsibility.

Even where, as in the Stoic idea of "sympathy," *sumpatheia*, there is an idea of cosmic resonance that extends well beyond the heavens and affects things on earth, including human bodies, that, too, is a matter of the order of things, not of human responsibilities.[80] The seamless whole formed by macrocosm and microcosms in China guaranteed the centrality of the emperor and the importance of advice to him (and of those who did the advising). Although Greeks too frequently modeled macrocosm and microcosms on one another, their notions of mutuality did not stretch to the idea that the behavior of human rulers was crucial to the welfare of the whole cosmos.

Greek philosophers devoted considerable effort to praising justice and censuring injustice in individuals in all walks of life. But the rational force that Greek teleology put in control of the cosmos was an impersonal one. Plato might hope that the philosopher-kings would exemplify the intelligence exhibited on a cosmic scale by the Craftsman in the *Timaeus*, but he did not think that a failure to produce philosopher-kings would inhibit that divine activity. The reward the philosophers looked for was not recognition of their crucial importance in achieving the welfare of "all under heaven"; rather, it was their own happiness, their understanding of how things are. As the Greek city-state declined from the Hellenistic period on, intervention in political life was increasingly represented as a potentially risky distraction from the essential aim, peace of mind.

Conclusions

Some would see the development of such notions as element, nature, cause, cosmos, as unproblematic, inevitably resulting from the internal dynamic of philosophical and scientific inquiries. Some would further argue that relating those concepts to the social, political, and institutional factors that we have invoked is misguided because it ignores or discounts the personal contributions of such geniuses as Plato or Aristotle.

As for the second objection, the factors we discussed did not *determine* Greek intellectual productions in all their variety. They formed part of the interacting manifold we have been exploring. Thinkers reacted to aspects of Greek legal experience sometimes positively, sometimes negatively. Again, which political models would be carried over into cosmology was far from a foregone conclusion. When we ask why this or that Greek philosopher or scientist proposed this or that idea at a particular historical juncture, neither purely social nor purely intellectual analysis can necessarily yield determinate answers. Our point is that an inclusive view is the best way of gaining an adequate understanding.

As for the view that there is no more to these developments than the internal dynamic of intellectual inquiry, we have stressed that traditions of philosophy and science evolved quite differently outside Greece. There is nothing natural, in the sense of inevitable, about the Greek conception of nature. Unless philosophy and science are defined as Greek philosophy and science, there is more to these problems than the automatic application of unclouded self-conscious reflection. Any attempt to privilege one mode of philosophy and science over any other is bound to be arbitrary.

The dynamics of the Greek developments we have explored in this chapter comprised three main dimensions: the circumstances within which Greek philosophers and scientists operated; the factors that related specifically to the domain of Greek legal practice; and those factors that were more strictly political in nature.

The suggestions we have made with regard to the last two areas are, in both cases, complex ones. It is uncontroversial that some of the primary Greek notions to do with causation and its connection with responsibility reflected legal experience. Our further argument relates to the pervasive phenomenon of adversarial debate in just

about every area of Greek philosophy and science. In the background
are models of such debates in the law courts—where large numbers
of Greek citizens personally experienced, both as litigants and as
dicasts, the modalities of prosecution and defense and the full gamut
of persuasive techniques.[81]

This influence is complex because although many thinkers
imitated forensic advocacy more or less self-consciously, others
treated such styles as models of what to avoid. For Plato and
Aristotle and those who followed them in demanding strict demonstration, mere persuasion would not do. That very suggestion
depended on the appropriation and transformation of the vocabulary of "proof" and "proving" that was familiar in the courts. They
represented demonstration as the opposite of mere persuasion, but,
as Aristotle recognized, demonstration was the most persuasive mode
of persuasion.

The political experience of Greek citizens was as varied as their
legal experience was extensive. Within the small-scale city-states of
the classical period, citizens not only contemplated, but in many
cases put into practice, radical solutions to the questions of
politics. The reasons for those political developments are complex,
stretching well beyond what we can discuss here. Their impact on
the formulation of the fundamental issues of Greek philosophy and
science was not limited to the direct application of a variety of
political models in cosmology. It is not just that cosmic order could
be represented in terms of the rule of one or many, or of no rule
but that of strife. The additional factor is the existence of alternatives.
Just as in the law courts prosecution was countered by defense, and
imputations of guilt by protestations of innocence, so, in the
political domain, one policy was pitted against another, not just on
what to do in this or that concrete situation but on constitutional
questions as well.

The extra role that Greek political experience thereby played was
that it offered precedents for the exploration of radically new

solutions to traditional problems. The influence was not one-way, from political life to philosophy and science, because the radical revisability entertained in the latter domains fed back into other areas of Greek experience. Plato and Aristotle were not the only philosophers who saw revolutionary new ideas in natural philosophy and science as potential threats to the good order that they insisted on in the moral and political domain. The paradox was that while both Plato and Aristotle saw stability as so desirable, they were locked in destabilizing disagreements not just with the opponents they shared but even with each other.

We come back, finally, to the circumstances in which Greek philosophers and scientists operated, where the key contrast with China lay in their comparative isolation from positions of political influence. The classical Greeks had no emperors to persuade; they had no sense of working toward an orthodox worldview that would at once legitimate and limit the emperor's authority, as well as bolster their own positions as his advisors. Even under the Roman empire, Greek intellectuals did not see it as their task to produce a cosmological underpinning for such a regime.

The Greeks made their reputations, rather, in what were often highly confrontational debates with rivals. What was at stake was not just fame but livelihood, insofar as that depended on attracting and holding pupils. The fundamental questions about elements, nature, causes, the relation between mathematics and physics, cosmos—each had its distinctive role to play in the management of those debates.

We asked why these questions became the focus of so much Greek philosophy and science. The answer, we suggest, lies at least partly in the ways those concepts were deployed in polemic. Each was used either to mark out a particular subject matter or to define a particular approach to it. Such moves were generally preemptive strikes in the controversy with opponents, whose positions were fatally flawed, from one's own point of view, either because they had no

idea of the real question or because their way of answering it was skewed. Even when there was general agreement as to the form of the theory needed to gain victory (for example, an element theory was needed to secure the foundations of physics, or a cosmology to give a comprehensive worldview), the disputes over the substantive answers reverberated on the understanding of the issues themselves.

It was not that all Greek discussion of these questions was merely the opportunistic exploitation of what appeared to be good debating points. But many participants paid due attention to what made for the strongest claims, and the concepts considered here played an important role in bids to secure a knockdown victory. This urge to defeat all rivals largely stimulated the development of these concepts in the first place.

In the process, much of what had been commonly assumed was challenged, and new notions of reality, some of which seemed counterintuitive to other Greeks themselves, were advanced instead. The open-endedness of the debate—with nothing immune to scrutiny—was, from one point of view, one of its undoubted strengths. Examining the foundations of beliefs led to the detection of possible inconsistencies (even if thinkers examined others' views with more energy than their own). Probing for axioms revealed the relations of dependence between parts of an investigation—even if a claim to the self-evidence of certain principles was, on occasion, merely a stimulus to rivals to try to undermine them. Yet with the rivalry that secured the open-endedness often went also a stridency in the claims to have delivered not just truth but the incontrovertibly true. Moreover, that stridency in turn evidently ruled out what the Chinese, for their part, learned to prize, namely, the sense of cooperative effort to find the common ground for a consensus.

5 The Fundamental Issues of the Chinese Sciences

The Problem

What questions came to be fundamental in China for exploring the physical world? In other words, how did Chinese identify and map what lay outside experience of social relations and of the self? The borders between these domains of perception were artifacts, shifting as the social consensus changed. An important question is what made the borders, kept them, and changed them. The Chinese who began to think abstractly about heaven and earth were not just so many isolated individuals. What assumptions did they share? What directions did they take? What led them to settle on a few fundamental concepts, such as yin-yang and the five phases? Were they, like their Greek counterparts, contesting the conventional wisdom? Did a coherent ordering of experience result from the diverse motives and interests of intellectuals?

This chapter begins where Chapter 2 ended. We move on to consider, first, what motivated individuals and collectivities to take up rational inquiry. We then reconstruct how Chinese with these motivations gradually formed and re-formed a view of the cosmos, up to the point where they knit it into synthetic writings. We examine critically a series of Chinese viewpoints that superficially resemble the Greek dichotomy of appearance and reality, in order to see what the Chinese ideas mean in their own circumstances. That prepares us to look closely at the notion of microcosms, the Greek counterparts of which we have already noted (p. 183). Finally, we examine the content of the sciences to see how they fit into the manifold of activity and thought that this book is about.

Let us begin, then, with what ancient Chinese tell us motivated their studies.

The Aims of Inquiry

In China, no less than in classical Greece, what people thought was inseparable from who they were and what they wanted. How to convince readers that the body is a little universe; what aspects of heaven and earth the body resembled; what consequences this association had for larger frames of understanding; how exploring the relation of cosmos and microcosm might affect one's own livelihood and standing—these questions interacted in the minds of those who pondered them.

Without doubt, a running conversation on the deepest questions of man's existence in the universe and in society was going on in China as elsewhere. But this was hardly something like an ongoing academic seminar that ignores exigencies outside the ivory tower. Intellectuals were aware of their rivals' positions and formed and stated their own partly in response, but they rarely resolved their disagreements through face-to-face confrontation, however vehement they might be. But because we are concerned with the circumstances of thought, we will again broaden the focus to include the social and political dimensions of their careers. We will pay particular attention to a neglected but notable topic: one-way discourse aimed at rulers.

Most humanist philosophers before the Han period sought, or lived on, patronage. By the Han, thinkers who had any hope of official status—that is, most of them—also wrote for rulers. The exceptions, early and late, do not invalidate our choice of emphasis.

Classical Chinese writers tended to see the study of heaven and earth as the result of a ruler's intentions. The account of the beginnings of astronomy most familiar to educated people through the ages comes from the "Institutions of the Emperor Yao," a document from between 350 and 300 B.C. that claimed to be much more ancient. Once this emperor of high antiquity had unified his realm, and prosperity had become the norm, "he commanded the Hsi and Ho [families of hereditary astronomers] reverently to follow august heaven, calculating and delineating the sun, moon, and other celestial bodies in order respectfully to grant the seasons to the people." After detailing the duty of the members of these families to conduct seasonal sky rituals, the monarch summed up their charge: "to take the 366 days and, by using intercalary months, to fix the seasons and to define the year. If you earnestly supervise all your functionaries, your achievements will be resplendent." The technical and ritual labors were meant, this document asserts, to add to the monarch's charisma—in this case, through his grant to his subjects of the new calendar that began each year—and thus guarantee a stable social order.

The ideology of medicine also tended to be centered on the monarchic will. In the first century A.D., at the beginning of the *Divine Pivot*, one of the two books that make up the *Inner Canon of the Yellow Emperor*, the ruler declares, "I treat the myriad people like my own children, nurture the hundred surnames [that is, the clans of the well-born], and receive their taxes. I am sad because they do not have enough and, on top of that, are subject to illness." The emperor's sympathetic response to their illness was to master the doctrines of medicine and compile a scripture on it.[1]

Physicians in private practice appealed to a different set of motivations, equally conventional. Chang Chi, in his preface to the *Discourse on Cold Damage and Miscellaneous Disorders*, surveys them when he complains about gentlemen who neglect the study of medicine, even though it would allow them, "above, to treat the illnesses of their lord and parents; below, to save the poor and lowly from calamity; and, in between, to protect their own bodies and lengthen their lives, to nourish their vitalities." These are ideals, but Chang adds that he wrote his book after epidemics killed a majority of his own clan.

The king of Huai-nan says that people write books "to provide a view of how the Way opens up or is impeded, so that people in later times can know the best course in carrying out plans and making choices." That lets them "settle their spirits, nurture their ch'i, and attain a state of perfect peace that will let them take delight in what heaven and earth have bestowed on them." He thus connects two dimensions of motive underlying inquiry—namely, what one is trying to understand and in what spiritual state one aspires to live.[2]

The physician Ch'un-yü I left altruism aside when he began his unique autobiographical account: "When I was young, I took delight in medicine."[3] Another aim was to get along in the world. Chinese authors, like their counterparts elsewhere, tended to comment on it only indirectly, as the earlier remarks on livelihood have already made clear.

The stress on moral significance and political relevance meant that practitioners did not gather data solely for practical purposes. The *Divine Husbandman's Materia Medica* of the Later Han period, for instance, was more than a collection of information on medicinal substances. Each was classified according to political hierarchy (see p. 232) and correlated with the cosmic rhythms of yin-yang and the five phases.[4]

Most Chinese philosophers, including those who studied the sciences, believed that there was more than one way to approach the same *tao*, the Way of the cosmos (p. 204). Cognitive understanding gained through induction and deduction, on the one hand, and the fruit of intuition, contemplation, insight, visualization, and allied nonrational means, on the other, were complementary. Study was one of several kinds of self-cultivation. It provided understanding and useful knowledge of the world (which was one aspect of the Way). The deeper aspect of reality (the nameless Way) is so subtle that one can penetrate to it only through noncognitive means.

The *Book of the King of Huai-nan* puts it cogently: "What the feet tread does not take up much space; but one depends on what one does not tread in order to walk at all. What the intellect knows is limited; one must depend on what it does not know in order to achieve illumination." The right balance of cultivation (of which study is but one kind) puts one in tune with the Way. It is not only humanists who state this principle; it turns up regularly in writings on mathematical astronomy and the other sciences.[5] The polymath Ts'ai Yung in A.D. 175 was not the first to affirm that empirical study cannot provide a full grasp of the cosmos: "The astronomical regularities are demanding in their subtlety, and we are far removed from the sages. Success and failure take their turns, and no technique can be correct forever. . . . The motions of the sun, moon, and planets vary in speed and in divergence from the mean; they cannot be treated as uniform. When the technical experts trace them through computation, they can do no more than accord with their own time."[6] This is a far cry from Greek rationalistic aggressiveness.

This sense of many kinds of knowing did not keep Chinese concepts from being physical, although objectivity did not become an issue. Going beyond the cognitive limits of inquiry enabled the seeker to embody, not merely to understand, the Way. Greek concepts were no less value-laden, but their originators consciously strove to separate objective and incontrovertible knowledge from other kinds. Because objectivity and certain proof were impossible in many domains of knowledge, as we have seen, Greek controversialists often bluffed (e.g., p. 155). Even then, they sought to override objections by claiming that their arguments were demonstrative.

What corresponds in China to the Greek authority of demonstration was the authority of sagely origin. Chinese, as we have seen, tend to trace the origins of a discipline to a charismatic revelation by a legendary monarch. The ancient classics, whether of statecraft or of mathematical astronomy, contained all possible wisdom. Their revelators recorded in them, in esoteric form, the fundamental patterns of knowledge and praxis. Membership in the right lineage and the right kinds of cultivation prepared one to comprehend the depths of the classics. Once one embodied these patterns, the spontaneous responsiveness and conscientious action of the sage took over.[7]

Scientific pursuits in China thus did not aim at stepwise approximations to an objective reality but at recovery of what the archaic sages already knew. What they knew was fundamental patterns rather than items of knowledge. Comprehending the ancient wisdom of the classics was onerous because in bad times—and the times were always bad—it was impossible by intellectual effort alone to comprehend fully what the sages had revealed. One had to be initiated by an exceptional master, receive a personal revelation from some more than human power, or break through by a prodigious effort of self-cultivation.

The Greeks were like the Chinese (and unlike most moderns) in that they did not confuse compiling data with knowing. In Hellenistic times emulation of a predecessor entered the picture, but only in certain coteries. Some Greeks shared the Chinese view of inquiry as spiritual self-cultivation, but their aim was not to recover the lost wisdom of a golden age.

The World That Concepts Describe

Creating worlds takes time. Over the six centuries we are studying, three universes evolved successively: the one reflected in policy discussions and interpretations of portents in the Warring States courts, the one that the early philosophers of the macrocosm tried out, and the one that matured in the cosmological syntheses of the first century B.C. The story of these three stages is so complicated that we will only summarize it here. Because it has not been put together previously, we detail our evidence and reasoning in the appendix (p. 253).

We can trace the beginnings of cosmology to records of the fifteenth century B.C. From that time on, what occupied the center of the universe and mediated human access to it was the person of the ruler. The shifts from feudatories of the Shang and early Chou dynasties to the mutually hostile kingdoms of the Warring States era to the centralized government of the Han period changed the character of that centrality but left it intact.

According to the earliest Chinese documents (around 1450 B.C.), the territory of allies surrounded the person and lands of the Shang ruler, and beyond the allies were the territories of other peoples. The king was also at the resonant center of a spiritual domain that included, among others, the high god Ti, various gods who embodied aspects of the physical world, and the king's divine ancestors, who gave him, and only him, access to that domain.[8] By 400 B.C. the Chou king, like his predecessors, held hereditary dominion (potentially, at least) over all known lands. After centuries of rapid change had made China a multistate system, his commands no longer had force. Former vassals now spoke and waged war as his protector (that was how they put it). Despite their power, they still depended on the rituals that only he could perform, which linked him via his ancestral line to the celestial order. It was not until 256 B.C. that the state of Ch'in dared to set him aside, eventually pushing beyond kingdom to empire. The Han empire that followed the Ch'in drew on all the trends of thought in its time to construct, on a basis secular as well as religious, a new monarch-centered cosmology.

In the three centuries before 400 B.C., local rulers aggrandizing their power sought their own access to the cosmic order, no longer by way of their ancestors but via the arcane knowledge of experts. Some, at least, came to rely on diviners, astrologers, military experts, physicians, and others who knew, each in his own niche, something about how humanity and heaven and earth interacted.[9] By 400 the scholar-cosmologists had not yet come on the scene, but the technologies of prognostication were spreading through society. Within a century, even minor officials were routinely using almanacs and divinatory apparatus to decide on propitious times for their tasks. The almanacs recently excavated from tombs of the early third century B.C. also cover activities of craftsmen, farmers, and traders and thus were also meant, directly or indirectly, for people in these groups.[10]

The purveyors of knowledge were trying out systems for analyzing affairs based on numerical categories—linkages based on sets of two, three, and a great many other numbers of agents, events, and relationships. When people thought of the action of drugs, they thought consistently of two types, heating and cooling, and so on. The habit of thinking in such categories originated, we believe, in the fixed sets of vessels, offerings, shapes, colors, recipients, and so on, that were mandatory in very early royal religious rituals. There was no reason by 400 to settle on one or two categories. The competition between different experts for the ear of a noble encouraged them to elaborate a given category rather than accept the conventional alternatives or a rival's definition. As the give-and-take continued, the experts accumulated a formidable array to choose from and modify to their liking. For instance, a good many quite diverse sets of five social entities, activities, and phenomena in part overlap and in part disagree.[11]

Between the mid-third and the late second centuries B.C., as part of the metamorphosis of Chinese culture, cosmological categories passed through two kinds of change. First, the rise of patronage in the late Warring States era stimulated a profusion of ideas designed to attract the attention of rulers. As clients frequently moved between courts, each a distinct aristocratic subculture, different conceptions and styles of presentation encountered one another. Second, from Lü Pu-wei on, a succession of scholars turned all the available resources, from court usage to classical learning to microcosmic speculation, toward a new double aim: providing polities with a cosmic basis and persuading rulers to entrust much of their authority to their bureaucracies. This bold attempt to tie the fortunes of the state to systematic thought mandated a kind of coherence and systematic use of concepts that was unnecessary earlier.

Adapting utilitarian speculations to universal goals was not as great a shift as it may seem. The philosophical lineages and their scholarship arose in the same educated social stratum from which many of the earlier diviners and other purveyors of knowledge came. Some astrologers and other experts before the Han were officials, but others were competing in the general patronage free-for-all, alongside the few who were offering ideas. Not surprisingly, the scholars saw that the experiments in correlation already going on could be useful both for understanding and for livelihood. The old notion that human society depended on six ch'i and five agents (wu-hsing; p. 255) was already metaphysical in its thrust. Supplemented with yin and yang, it provided philosophers with a tight, essential linkage between society and the cosmos, missing in Confucius and his humanist successors but pointing toward new vistas of thought.

The untranslatable term ch'i was used before 300 B.C. for a multitude of phenomena: air, breath, smoke, mist, fog, the shades of the dead, cloud forms, more or less everything that is perceptible but intangible; the physical vitalities, whether inborn or derived from food and breath; cosmic forces and climatic influences (p. 256) that affect health; and groupings of seasons, flavors, colors, musical modes, and much else. Ch'i could be benign and protective, as that proper to the human body was, or pathological, an intangible agent of disease.[12]

Yin and yang between the third and first centuries B.C. became highly abstract—paired, complementary divisions for any configuration in space or process in time. They were equally applicable to discussing the interaction of active and reactive, growing and dwindling, masculine and feminine.

The five phases emerged as an analogous set of fivefold divisions, also complementary, of configurations or processes. "Phase" is an elegant English counterpart of the *hsing* in *wu-hsing*, a concept that occurs in the *Springs and Autumns of Master Lü* and thereafter. It reflects the common, nontechnical sense of "phase": "any one aspect of a thing of varying aspects; a state or stage of change or development." Before the late third century B.C., *wu-hsing* was much vaguer, sometimes best translated as "five activities" and sometimes as "five agents" (see the appendix).

One could apply either the five phases or yin-yang to a given complex phenomenon, depending on how elaborate an analysis one wanted. Thinkers from Lü Pu-wei on arrayed the five phases in various sequences, especially the production cycle, used to model processes of evolution and physiology, and the conquest cycle, used when one agent overcame another and generally for pathological interactions. Those and other sequences could model almost any mode of activity, both human and in the external world.[13]

The cosmologists of the third and second centuries B.C. settled on *ch'i*, *wu-hsing*, and yin-yang for reasons that were in part arbitrary, but these categories were prominent among those already circulating among the palaces. Ch'i would have attracted thinkers seeking an already broad concept and willing to broaden its sphere of reference even further. Yin-yang was adaptable to analyzing the complementarity usual even in very early Chinese thought. The fivefold analysis of the phases was no more or less natural than the fourfold rubrics that play a large role in Greek physical thought, or the triples used for other purposes in China. More than that we cannot say.

Lü Pu-wei, Liu An, and Tung Chung-shu are paramount among those who systematically used ch'i, the five phases, and yin-yang as a foundation on which to build philosophical doctrines of the cosmos, the state, and the body. Their writings contain hesitant and inconsistent but in the long run influential moves toward defining relationships between ch'i and the other two concepts. These moves succeeded only at the next stage, in the first century B.C. But the three authors' focus on these concepts out of all the current ones, and their linkage of ch'i with yin-yang, made the third stage of mature synthesis possible (see the appendix). Other scholars, deeply versed in divination, enriched these doctrines with new concepts and frames of analysis based on studies of the ancient Book of Changes.

A fully developed cosmological doctrine, in which yin-yang and the five phases became categories of ch'i, tools for analyzing its complex configurations and processes, appeared in the first century B.C. Detailed and generally compatible accounts appear in two books, Yang Hsiung's Supreme Mystery of around 4 B.C., a poetic meditation on time-bound change, and the anonymous medical Inner Canon of the Yellow Emperor, possibly earlier or later. That these books are so different in character is a reminder of the broad and continued philosophical effort that underlay both.

The Supreme Mystery and the Inner Canon set out a notion of the physical universe, drawing on many components that had been worked out earlier. According to them, the Chinese cosmos is a constant flux of transformation, always regenerating itself as its constituents spontaneously change. Ch'i is matter, transformative matter, always matter of a particular kind, matter that incorporates vitality.

Yin-yang and the five phases had, by the end of the first century B.C., a consistent, dynamic character as part of the ch'i complex. Anything composed of or energized by ch'i is yin or yang not absolutely but with reference to some aspect of a pair to which it belonged and in relation to the other member. An old man might be yin with respect to a young woman if it was a matter of stamina, or yang with respect to a young man if the topic was political power. Yin-yang provided a flexible language well suited to discussing the balance of opposites. This was a balance not of quantity but of the dynamic quality of each in interacting domains—for instance, something could be yang in its activity and yin in its receptivity. When the focus was not on a binary opposition, however, but on more complex sequences of growth and decay or conquest and subjugation within a larger process, the various sequences of the five phases came readily into play.[14]

The unification of ch'i, yin-yang, and the five phases was technical in character. The *Inner Canon* was not easy to master, because it jumbled together many short writings that contradicted each other in various details (even on the number of visceral systems). Still, later medical treatises gave systematic accounts of the classic, just as commentaries from the third century B.C. on found coherent ancient wisdom in the *Book of Changes*.[15]

The world was driven by the ceaseless dynamic of ch'i, which could be understood by analyzing it into two or five phases and investigating the resonance between things that belonged to the same category. This conception became prevalent among gentlemen who read the *Changes*, the *Supreme Mystery*, and the *Inner Canon*. It became indispensable because the raison d'être of the bureaucracy for which they worked (or wanted to work) was to cope with constant mutation by finding its inner order.

The main characteristics of cosmological thought show some interesting similarities to, and contrasts with, the situation in ancient Greece.

The Chinese had no reason to seek a counterpart to the Greek *phusis*. We have looked at a very few of the technical terms that early Chinese used when they defined the characteristics of animals, minerals, and plants, of heaven and earth, of the patterns of things, of spontaneous changes. What the various terms refer to sometimes overlaps certain denotations in English of "nature" and words allied to it. But before modern times, Chinese did not need a word that meant "nature" (the physical or material universe). The classical word most commonly so translated, tzu-jan, meant simply "something that exists or is the case (jan) without something else causing it (tzu)." Its usage broadened slightly over the centuries, but it did not mean "nature" until 1881, when Chinese borrowed this sense from Japanese modernizers who had introduced it to translate occidental textbooks.[16]

What concepts, then, did the Chinese invent to describe the domain of experience outside the self, society, and the realm of the gods? Historical analysis of ch'i, yin-yang, and the five phases supplies the main components of the answer as it applies to science and medicine. The notion of the body as microcosm gave medicine a focal role in the ch'i complex.

An equally important term in philosophy was "the Way," tao. Tao and ch'i took on complementary meanings in Han writing and remained central in cosmology and cosmogony—as well as in religious thought. Neither became an exclusively physical concept. The notion of a purely physical concept did not attract Chinese—nor was it requisite for sophistication in scientific thought.

Tao and ch'i belong to different levels of abstraction, tao to the highest. It began as simply a word for "road" or "path," but in the hands of Confucius and those who followed him it took on normative meanings. Tao is the proper path in life, the one the sages follow spontaneously and others strive to follow.[17] It is, at the same time, Lao-tzu's mystical ground of process (not of being). Your way is not who you are but what you do, not the species of a tree but how it grows. The Book of Chuang-tzu imports this notion into philosophy with a story in which the tao of a robber is not his deviance but the skill with which he loots your house (p. 70).[18]

This understanding was not the property of Confucius, Lao-tzu, Chuang-tzu, or anyone else; classical philosophers shared it. Philosophical and scientific collectivities acknowledged the one great Way, an organic *tao* that interweaves the individual ways of everything in the universe. For most, but not all, thinkers, it is also the Way of human society, connecting the life trajectories of individuals. *Tao* is more prominent in everyday thought than in science, technology, and medicine, but it is the overarching concept in what, as any comparison will confirm, is a philosophy of process.

An eloquent portrayal of the Way and its role in the good life and the good state became a trustworthy means to engage the attention of rulers. Philosophers who spoke for any such view could use it to compete for appointments as Erudites or in other posts. In none of the jockeying for support (and vying to say what the ruler wanted to hear) did competitors draw attention to the many discrepant but equally estimable Ways in the teachings of diverse masters; to the contrary, they were more likely to level charges of heterodoxy. Nor did they (even those without political ambitions) suggest that it would be good for numerous Ways to coexist. All agreed that there was only one. The issue was whether someone was right who claimed that his description of it was the true one. The synthetic writings of the last three centuries B.C. gradually redefined *tao* to identify it with the new unified, centralized empire and its foundations in the cosmic order. After the Han period, promoters of organized Taoist and Buddhist religious movements further elaborated this political role of the Way, borrowing back and forth with discreet abandon to compete for the support that Confucian teachings had largely lost.

Han thinkers made ch'i the material and energetic basis of things and their transformations, and the five phases and yin-yang, once and for all, aspects of it. In the sciences, this ensemble came into its own. Ch'i bridged the transition from humanistic thought to state cosmology and then to distinct physical sciences and, in doing so, kept the latter politically subservient.

The mature cosmological synthesis of the late first century B.C. evolved, then, out of state-centered efforts to combine the cosmic, sociopolitical, and somatic good in the late Warring States period. The synthesis began with officials and clients creating new rationales for advice to rulers in a rapidly changing multistate world. Their efforts converged in the third and second centuries B.C. First, philosophers within the patronage system evolved a systematic language that served their own intellectual concerns and argued for influence over affairs of state. Next, Lü Pu-wei and his successors attempted to make the emperor a ritualist immersed in self-cultivation, withdrawn from the daily work of management, drawing on and freely reinterpreting many philosophical currents.[19] They perfected a common vocabulary to describe cosmic process and its analogues in the state and the body. The result was a comprehensive rationale for government and for a new separation of powers within it.

The result in the long term was not exactly a victory for the bureaucracy and its ideologists. They were far from the top of the official hierarchy. Given their lack of power to compel the emperor and his closest advisors to do anything, the influence of their vision of monarchy and the civil service is impressive. In practice as well, their point of view became part of court discussions. A very few emperors were even willing to reign more or less in accord with their visions. The Extensive Emperor (r. 73–49 B.C.), for instance, was known for his reluctance to interfere in administration, his personal participation in state rituals, his abstemiousness, and his reservations about conventional moralism as a basis for government.[20]

But officials paid for their enhanced authority in the currency they had minted. They had recommended that the monarch be a distant figure involved primarily in self-cultivation, but they still pictured him as the ultimate source of orders that officialdom promulgated.

The outcome was a more rational system than it might otherwise have been. Still, a determined sovereign could force them, sometimes easily and sometimes with great difficulty, into subservience. Because their status was official, the discussions were increasingly one-way. Thinkers proposed; the ruler or his surrogates disposed.

Appearance versus Reality

Having examined the history of a complex of ideas unique to China, we are now prepared to investigate how and why Chinese conceptions, despite superficial likenesses, differed from ideas important in the Greek world. Our example will be the contrast, prefigured in Homer and Hesiod and influential from Heraclitus and Parmenides on, between appearance and reality (pp. 142, 175). Many claimed that the causes of the phenomena that human beings experience through their senses can be found on levels accessible only to reason. In other words, reality is hidden; as Democritus put it, the richness of flavor and other sensory perceptions are nothing more than the motion of atoms in the void. But the underlying reality is no less physical than the phenomena it explains.

Ancient Chinese found no reason to doubt that the fundamental physical realities were what they could see and touch. Possessors of the Way were not motivated to reject common sense and base the physical cosmos on a hidden order of things. In this they were unlike Masters of Truth in the Greek world, looking for fresh ways to trump their rivals. Appearance versus reality became a Chinese issue only with the introduction of Indian metaphysics, which first made a splash in the third century A.D. But that was spiritual, not physical, reality.[21] In our own period, four distinct stances, epistemological and polemical, are (among others) each analogous in its own way to Greek notions. These are the contrast between an accessible Way and an ineffable one; assertions about the qualities that make some specialists better than others; the distinction between empty and full;

and advice on detecting spurious resemblances. They turn out to have little in common with the fundamental Greek dichotomy. Local circumstances account for the differences.

The *Book of Lao-tzu* begins with **two Ways** (*tao*). There is the one that can be spoken of, that was the mother of the myriad creatures but is not the constant Way, and there is the nameless one, the beginning of heaven and earth, the Way that is constant.

> These two are the same
> But diverge in name as they issue forth.
> Being the same, they are called the mystery,
> Mystery beyond mystery
> Gateway of the manifold arcana.

This is not meant to be natural philosophy. Nor does it claim that only one of the two ways is real. Nor, for that matter, is its topic cognition. It is about the difference between common experience and mystical breakthroughs. Because the deeper Way is mysterious, arcane, and subtle, a sage cannot give an orderly account of it to the unenlightened without lapsing into paradox.[22]

Roughly a generation later, the *Springs and Autumns of Master Lü* turned the notion of the sage as the ideal ruler inside out to persuade the ruler to become a sage. A section entitled "Investigating Subtleties" begins: "If order and chaos, survival and extinction, stood in the same relation as a high mountain and a deep valley, or as white clay and black lacquer, there would be no need for wisdom; even fools would do. But order and chaos, survival and extinction, are not that way. They seem knowable, and then not; they seem perceptible, and then not. . . . The beginnings of order and chaos, survival and extinction, are like the fur of autumn. If we investigate autumn fur, we will not blunder when it comes to the big things."[23] "Autumn fur" is an animal's almost imperceptible downy hairs that by winter develop into new fur, a common metaphor for the subtle first stirrings of change that only a sage detects.

The concern of this passage is with social disorder, which begins small but grows quickly and disastrously if it is not dealt with. The order and chaos, survival and extinction, are those of states. The investigation is a practical matter of avoiding nasty surprises.

It is easy to find many similar passages in writings shortly before and during the Han. The issue is never the unreliability of experience.

Those whose livelihood comes from being able to predict the future or to determine what has gone wrong in the human body must satisfy their clients that they have access to special **knowledge** not open to everyone else. Chinese diviners and physicians did not generally stake their authority on the metaphysical foundations of what they knew, nor on the formal rigor with which they presented it. Unlike the Greeks, they did not contrast experts' hidden realities with ordinary people's appearances. Their qualifications tended instead to be social. Because expertise was not inherently problematic, it was initiation that separated insiders and outsiders, and gentlemanly behavior that marked the superior insider.

Experts' claims to superiority had two main components. The first was to trace the knowledge of their occupational group back to the legendary sage emperors who had originated culture and granted it to their subjects. For instance, the profundity of medical as well as astronomical learning was guaranteed by the Yellow Emperor, who created and revealed the first technical classics, and the Divine Husbandman, who tested and set down the healing virtues of medicinal plants.

The second component was a lineage of textual transmission (see p. 58). The chain of orthodox masters and their disciples directly linked the learner to the original revelation. This linkage was necessary because the founding classics were too profound, scholars believed, to be understood outside that line. The imperial physician Wang Hsi, in his *Canon of the Pulsating Vessels*, written just after the end of the Han period, emphasizes the obscurity of the ancient books.

"Through the ages few have been able to draw on the extensive meanings of the writings that survive. The secret implications of the old classics have been kept arcane rather than been broadcast. This has left scholars of later eras in the dark about their fundamental meaning, each with his own partial view, unduly confident of his abilities. The result is obvious: minor illnesses transformed into life-threatening ones, and chronic problems dragging on until all hope of recovery is lost." Those who through study "tread in the footsteps of the ancient worthies can avoid causing premature deaths." The footsteps of the ancient worthies, perceptible in their writings, lead back to the founding revelations of medicine and guarantee the efficacy of treatment.[24]

This approach to justifying expertise was visible earlier in astrology. The Grand Scribe Ssu-ma Ch'ien explains its origins: "Long before the time of the Divine Husbandman, it would seem, the Yellow Emperor determined the paths of the heavenly bodies, established the motions of the five planets, began tracking the variations in the celestial motions, and corrected the Intercalation Remainder [used to add a thirteenth lunar month to years at regular intervals]. From then on, there have been officials in charge of heaven and earth, of the gods, and of the various categories of things. These we call the Five Officials. Each is responsible for maintaining the order [of his charge] so as to avoid disorder." The bureaucratic character of the art in Ssu-ma's own time, he asserts, is needed to support the archaic emperor's revelation.[25]

What earned favor for certain astrologers and diviners before the Han? Much early evidence lies in *Master Tso's Tradition of Interpretation of the Spring and Autumn Annals*. The answer is neither arcane knowledge nor empirical efficacy.

This is clear enough in the case of the best-known astrologer of the sixth century B.C., Ts'ai Mo. The *Tso Tradition* records with obvious esteem his interpretations and prognostications regarding affairs of

state over twenty-six years. His noble employers, the ultimate decision makers, greeted every one of them with silence.[26]

What, then, made Ts'ai paramount among diviners? The king of
Wu, aware of Ts'ai's reputation, asked an envoy, "How did Ts'ai come
to be considered a lordly man (*chün-tzu*)?" The answer was, "Ts'ai,
when he put himself forward, incurred no dislike, and when he took
his leave, was not criticized." The king replied, "His reputation was
deserved." To paraphrase, Ts'ai was a gentleman in the best conventional mold, a modest and faithful civil servant.[27]

Han histories have more to say about the service of diviners and
other technical experts (who had been on the scene all along). One
story, reviewed earlier, was about a street diviner whose gentlemanly
qualities embarrassed two high officials (p. 21). On the other hand,
some rulers of the time were aware that the customary inheritance
of technical posts provides scope for incompetence: "Although the
hereditary posts passed from father to son one generation after
another, much of the diviners' subtle mastery and deep knack of
interpretation was lost." What brought this dangerous situation to a
head? "When the Martial Emperor was enthroned, he opened
wide the road for those who had mastered arts and skills, inviting
[to his court] practitioners of a hundred kinds of studies. Every gentleman who had mastered a skill had an opportunity to demonstrate
it. Those who were the best of their kind, outstandingly impressive,
were given posts to assist him. In the course of several years he
assembled a number of imperial diviners. At the time the emperor
wanted to attack the Hsiung-nu, repel the people of Ferghana west-
ward, and absorb the Yüeh peoples to the south." In other words,
because the Martial Emperor considered divination essential to his
expansive military ambitions, he rebounded from the hereditary
principle to that of merit. As usual, political authority, responding to
its own exigencies, made fundamental decisions about technical
qualification.[28]

Practice outside officialdom was another matter. Although historians know much less about it, the sources paint a consistent picture of authority rooted in transmission. It is clear from the report of the renowned physician Ch'un-yü I (p. 77) that a doctor's lineage and the books that had been transmitted to him were the most important components of technical qualification as the government defined it.

To sum up, in the view of those who paid for technical specialists, what made them outstanding was not the conceptual basis or content of their knowledge, an exceptional intellectual grasp of fact or truth. It was specifically the chain of predecessors, all Possessors of the Way, that connected them directly and intuitively to the kingly revelations of high antiquity. What guaranteed "subtle mastery" was these origins and the successful transmission of the tradition down to them. Their patrons and employers largely took their problem-solving skills for granted, except when faced with incompetence that proved flawed transmission. Comportment appropriate to the functionary's situation in life was a sign of competence. Their superiors expected them to behave like gentlemen.

The labels "**empty**" **and** "**full**" (*hsu-shih*) imply not only the value of something but also what knowledge one can have of it. The meaning of this dichotomy, like that of other complementary pairs such as yin and yang, varied with the circumstances and the topic. An essay entitled "Five Officials" (probably third century B.C.) that found its way into the *Book of Kuan-tzu* discusses the keys to victory in war. Among them is "differentiating real from empty," that is, recognizing the difference between the actual situation and disinformation, ruse, or trap. But when another writing of about the same time in the same collection discusses the ruler's "discipline of the heart and mind," emptiness (that is, void) is an attribute of the sage, which sets him off from ordinary people, who are preoccupied with fruitless bustle: "The sage does not do this, so he differs from other creatures. In his difference he is empty; the empty is the starting point of the myriad things." He is the still, generative center in the midst of meaningless change.[29] Both meanings, though divergent, persisted over centuries.

But when the topic was ideas or words, the empty ones are consistently inferior. The quietist authors of the *Book of Lao-tzu* esteemed emptiness and other yin attributes of people and things, but that was not true when they spoke of *hsu yen*, "empty words or empty sayings": "When the ancients said [of the sage that he may be] 'bowed down but is intact,' surely these were not empty words." That usage is normal in scientific and medical writing as well. The author of the *Divine Pivot*, for instance, sketches the character of people who are endowed at birth with a strong imbalance in the direction of yang *ch'i*: "They always feel at ease and love to talk of grand enterprises. They lack ability, and their words are empty."[30]

"Empty words" is a tool of rivalry. It asserts that in what an opponent has said there is nothing worth taking seriously. Your quarry will not agree with you, at least if he is on hand to defend himself (which is unlikely to be the case). That there may be no impartial way to settle the argument does not matter, however. You have had a crack at discrediting him and have done so in a way that is not itself obviously empty of substance.

This controversial application of "empty versus full" crudely parallels the Greek use of "appearance versus reality." But their consequences for thought are quite different.

The Greeks were generally explicit and confrontational when they pressed arguments against rivals. Because claims to possess sure knowledge were common currency, both sides in a given debate tended to elaborate and push as far as possible their own notions about reality. It is difficult to ignore a charge that your reality is mere seeming, and tempting to vindicate it by demonstrating that the confusion is actually your opponent's. One bold stroke in critical thought after another resulted from such contentions.

The Chinese dichotomy evolved for equally impolite uses in quite dissimilar circumstances. One could apply the label "empty" to more or less any assertion, with no more epistemological significance than calling it "shabby" or "silly." It normally amounted to nothing more than a charge of dimness or wrongheadedness. The point was not to launch critical arguments but to end them. This usage could have been developed in philosophical directions, just as, after the Han period, the other usage of *hsu* as "void" came to play an important role in Taoist and then Buddhist metaphysics. But this did not happen. Han Chinese found the vituperative sense of "empty" adequate for dismissing the claims of rivals and phrased their speculations about the character of knowledge in other terms that were not so openly disputatious.

Overt, reciprocal polemic of a kind that might have pushed epistemological problems to the fore was rare. *Chih*, the staple verb for "to know," overlapped in philosophical writing with words for "to recognize" and "to know how" and, as a noun for "knowledge," with "empathy" and "wisdom." Most authors interested in epistemological matters found no reason to draw a rigid line between "wise" and "knowledgeable," between those who understand and those who use information effectively.[31]

One more instance, an odd but engaging one, will help explain why Chinese thinkers of our period, though not generally fascinated by epistemological issues, found the notion of "empty" so useful. The *Springs and Autumns of Master Lü* contains a most interesting chapter on **"spurious resemblances"** (i *ssu*).

What most confuses people is surely resemblances between things. What bothers jade cutters is stones that resemble jade; what bothers judges of swords is swords that look like [the legendary blade] Kan-chiang; what bothers worthy rulers is people who know so many things and quibble so well over words that they seem to be learned. The ruler of a doomed

state may seem to be wise; the ministers of a doomed state
may seem to be faithful. Resemblances between things
greatly confuse the stupid but cause the sage to reflect more
deeply. . . .
  It is essential to examine what lies behind spurious
resemblances. Doing so depends on [access to] the right
person. If [the legendary king] Shun were a charioteer, with
Yao seated at his left and Yü at his right, before entering
marshland they would consult a herdboy; before crossing a
stream they would consult a master fisherman. Why is this?
Because what they needed would be thorough knowledge. A
mother can always distinguish identical twins because of her
thorough knowledge of them.[32]

One can distinguish true from imitation in many ways; here the
issue is valuable things versus things of inferior worth that merely
resemble them. One distinguishes them, it turns out, by consulting
experienced people!

  Yao, Shun, and Yü were sages, the ultimate brain trust, certified
by well-known classics. Still, readers did not need to be convinced
that the three would need the advice of a herdboy to find their
way through a swamp. Nor would they doubt that experience is
pertinent to detecting phonies. The author did not find the notion
of experience problematic. Neither did contemporaries or later
scholars.

  Master Lü seems, at a glance, to offer a curiously trivial resolu-
tion for the substantial problem of real versus specious. But from his
viewpoint, such a solution matters.

  Rulers, the readership for whom intellectuals yearned, generally
were not enthusiastic about being subjected to rational suasion.
Rulers from Confucius's time to the end of the Warring States era,
no matter how respectfully intellectuals spoke to them, usually
appear in philosophical writings as well-intentioned dolts, their

attention spans minimized by their appetites and their power. Advisors trying to persuade them that a given policy was in their interest were as likely to annoy as to compel them. Still, if one hoped to be an advisor, it was them, not some imaginary paragon, one had to advise.

Every essay in the *Springs and Autumns of Master Lü*, as in most other books of our period, promoted stability imposed by the state. The passage just quoted was not epistemological in aim. It speaks to the decision maker who may nod off when confronted with the dilemmas of jade cutters but who takes a lively interest in detecting advisors who pretend to be well informed. The emphasis of this and other chapters on trusting experienced advisors was eminently practical. Such parables might forfeit that practicality if they set civil servants arguing over whose knowledge was most like that of the mother of twins—in other words, over what experience counted and what did not. A main aim of the book, after all, was to persuade the ruler to cede authority voluntarily to his officials rather than acting unilaterally. The choice was his. Making it attractive necessitated uniting officials rather than setting them at each other's throats.

This example from Master Lü throws some light on why Chinese thinkers used the notion of "empty" to dismiss the ideas of rivals and felt no need to draw out its epistemological implications. When Greek Masters of Truth argued, the audience was all or some subset of their fellow citizens. Shortly before the Han, Possessors of the Way were hoping that aspirants to ultimate power would read and appreciate them. In the Han era they were writing for the emperor, the Son of Heaven, and those who acted in his name. Consciousness of these Chinese readerships encouraged precision in moral, social, and political categories, but it did not motivate an equal fastidiousness with regard to the foundations of knowledge, even when discussing abstractions. For those who sought a justifiable claim to knowledge, revelation by an archaic sage-king provided an intellectually impeccable one that monarchs might appreciate. It countered the charge of officials that pedants without hands-on experience could offer no advice of any value.[33]

That reliance on a tradition may well explain why the most promising foundational explorations came from an unconventional organization in the late Warring States era, the Mohists. The *Mohist Canons* of the late fourth century B.C., included in the *Book of Mo-tzu*, opened up systematically a range of epistemological issues, from the difference between knowledge and wisdom to the relations between names and things, not to mention unprecedented explorations of optics, mechanics, and basic geometric topics.[34] But the independence of the Mohists also accounts for the abortiveness of their thought. Their authoritarianism may have pleased rulers, but their preaching of impartial love for everyone (undermining the claims of hierarchy and kinship), their opposition to elaborate ceremony, their emphasis on frugality, their opposition to offensive warfare, and their disruption of military conquests predictably offended those who were winning. In the fourth century B.C. some considered them the chief rivals of the ju as the claimants to moral authority. But they did not benefit notably from patronage and did not long survive what after 136 B.C. became the ju monopoly of state classical studies.

All these examples are reminders that direct experience reigned over early Chinese notions of external reality. A strategy that succeeded in the competitive hurly-burly of the Hellenic world did not attract Chinese thinkers whose concern was first and foremost persuading a ruler or his surrogates to want their advice. Intellectuals established through membership in lineages of learning and through conduct befitting civil servants the authority that Greeks asserted with claims about rigorous demonstration. For Chinese cosmologists and scientists, the world of sight and touch remained a source of knowledge that was quite reliable for their purposes. The patterns of argument, including techniques for reproaching, parrying, or ridiculing opponents, reflected the circumstances and milieux in which intellectual exchanges took place.

Macrocosm and Microcosms

Although several Chinese conceptions superficially resemble the Greek antinomy of reality versus appearance, it is now clear that they differ in fundamental ways and that the differences in their meaning and in their use are inseparable. We will now examine an idea that does turn up in both Greece and China. Its content is broadly similar in both, but important differences call for explanation.

A basic feature of systematic thought about the external world as it arose in China is that the body and the state were miniature versions (not just models) of the cosmos. By the late first century B.C., various thinkers had invented authoritative forms of the three realms. Links between the three evolved much earlier, but by this time the connections were systematic and tight. The creators of these systems were preoccupied with political authority and its effective use. They made the emperor the indispensable mediator of this set of mutually resonant systems. As we have seen, in doing so they transformed him, to the limited extent that they could, from a wielder of raw power into a sagely ritualist.

From the first historic dynasty (ca. 1570) on, Chinese, like Greeks, reinvented the state again and again. Although both peoples drew on the experience of independent local governments, the Chinese did not invent diverse constitutions. Governmental institutions varied in interesting ways from one state to another, but predatory consolidation eventually wiped out all but traces of the diversity.

After China was united in 221 B.C., the functions of the government changed greatly over the two millennia that the empire lasted. They varied with the size of its dominions, the technology and effective span of its control, its budget, the relationships of emperor and officials, and the social origins of civil servants. But Chinese persisted century after century in imagining the state as an unchanging authority, to which the experiences of one or two thousand years earlier were directly applicable.

The **cosmic order** that Chinese imagined also differed greatly from that of the Greeks. Like the functionaries of Mesopotamia before them, those of early China believed that irregularities were ominous, meant by heaven to warn rulers. The Greeks did not build their astronomical models atop this conviction, although they borrowed much else from the Middle East. Chinese worked out their cosmos without significant borrowing.[35]

From 200 B.C. on, a corps of astronomical experts, entrenched in a large imperial civil service, used standardized procedures to scrutinize sky and earth for omens joyous and baleful. Their work was essential to maintaining the mandate of the ruling house. They had to be alert when heaven intervened with portents. Their reports warned the ruler when he flagged in his endless tuning of the state's activities (religious as well as strategic) to match the rhythms of heaven and earth.

Well into the Han period, one of the interpretative traditions of the *Book of Changes* expressed this relation in its metaphor of the three powers (*san ts'ai*). Man, symbolized by the central pair of lines in each hexagram and personified in the ruler, stands intermediate between sky and earth (in the hexagram on p. 267 in the appendix, the upper and lower pairs).[36] This pattern was not widely accepted before the Han. Confucius and his successors were convinced that establishing the good life was a problem to be solved within society.[37] As we have seen, the ideological syntheses of the Han swamped Confucian humanism in cosmology.

The classics sponsored by the Han government, in asserting that sage rulers in high antiquity associated the state with the order of sky and earth, provided the ultimate precedent for microcosmic thought. The "Great Plan" (probably written between the early fourth and the mid-third centuries B.C.; p. 259), a short text passed down as part of the *Book of Documents*, had already connected the two spheres with a broad array of correspondences. They include many numerical categories, mostly fivefold, although most of these did not shape later concepts. A passage speaks of "the year, the month, the day, the stars and constellations, and the calendric calculations" as the "five regulators." The text is much concerned with divination and celestial portents, which it segregates according to rank: "The king watches the Year Star, Jupiter, [for portents]. Ministers and officials watch the moon. The leader of the army watches the sun. . . . The common people watch the stars."

The explicit purpose of the "Great Plan" was to guide a king and to support his authority through ritual, teaching him how to manipulate the elaborate, ordered array of correspondences that heaven originally granted to an archaic ruler. The document aims not only to root out but to prevent opposition: "Whenever among their people there are no depraved factions and no rival powers, it is precisely because the great [kings] have created the standard."[38]

The understanding of heaven and earth and their relation to humanity evolved alongside the definition of the state and rulership, in fact in the same documents. As the king of Huai-nan put it in the second century B.C., "Heaven displayed the sun and moon, arrayed the stars and the other markers of time, adjusted the balance of yin and yang, spread out the four seasons, with day to expose [all creatures to light] and night to give them rest, wind to dry them, and rain and dew to moisten them. As heaven gives birth to the myriad creatures, no one can see how it nurtures them, but they grow; as it kills them, no one can see how it kills them, but they die. This is what we mean by its divinity. The sage models himself on it, so that as he engenders felicity, one cannot see how, but felicity arises; as he roots out calamity, one cannot see how, but calamity goes away." Earlier in the same century a royal advisor socially segregated the modes of emulation he recommended, in much the same way that the "Great Plan" did, by watching for portents: "The ruler models himself on heaven; his helpers model themselves on Earth; his assistant ministers model themselves on the four seasons; the people model themselves on the myriad phenomena."[39]

Sage rulers thus created the institutions of the empire to echo the order of heaven and earth. In doing so they defined the macrocosm-microcosm relation. That is the official account. Hindsight suggests, rather, that as rulers and their intellectuals fashioned the political order, they simultaneously projected it outward to define a cosmos. How did they fashion these parallel universes? Surveying the many permutations of correspondence will answer this question.

Lü Pu-wei, planning a future empire, eloquently envisions **the state as cosmos**:

The Way of heaven is round; the Way of earth is square. The sage kings took this as their model, basing on it [the distinction between] above and below. How do we explain the roundness of the Way of heaven? The essential *ch'i* alternately rises and falls, completing a cycle and beginning again, delayed by nothing; that is why we speak of the Way of heaven as round. How do we explain the squareness of the Way of earth? The myriad things are distinct in category and shape. Each has its separate responsibility and cannot carry out that of another; that is why one speaks of the Way of earth as square. When the ruler grasps the round and his ministers keep to the square, so that round and square are not interchanged, his state prospers. . . .

The One (the ineffable Way) is most exalted of all. No one knows its source. No one knows its incipient form. No one knows its beginning. No one knows its end. Still, the myriad things take it as their progenitor. The sage-kings took it as their model in order to perfect their natures, to settle their vital forces, and to form their commands.

A command issues from the ruler's mouth. Those in official positions receive it and carry it out, never resting day or night. Moving unimpeded all the way down, it permeates the people's hearts and propagates to the four quarters [of

the realm]. Completing the circle, it reverts to the place of
the ruler. That is the Round Way.[40]

In this passage the cosmic Way itself mandates strictly separating
the responsibilities of ruler and officials. The monarch's obligation
is to cultivate the Way so that his commands, like his rituals, are in
harmony with it and therefore potent. Although he is not passive,
his own will is no more involved in executing commands than in
forming them. If he is a sage, his mental and physical activities are
spontaneous. It is the duty of his officials to make the verbal embodiment of his command circulate throughout the realm and then back
to the court, exactly as vital ch'i circulates through the human body
(cf. below, p. 223). The customary minute division of the civil service
into posts with distinct responsibilities is as firmly grounded in the
cosmos as is the ruler's sagehood. "Separate responsibility (fen-chih)"
is also what distinguishes the myriad things in that great bureaucracy
the cosmos. The spontaneously orderly interaction of everything in
the universe, not the imposed will of a creating or ruling deity, estab-
lishes these posts. In this sense the political domain and the universe
do not differ. The emperor, his subordinates, and the rest of his realm
constitute a spontaneous organism—a microcosm—held together by
his rule. His commands are necessary, but can be successful only if
they, too, are spontaneous and in concord with the larger ensemble
of processes.

Ancient peoples perceived the interior of **the living body as a
cosmos**, combining cognitive ingredients, social ideals, physical
data, and sensual self-awareness. Even vivisection (p. 98) could reveal
little about the processes that keep the body alive. Most Greeks imagined it primarily in terms of structures, organs, tissues, and liquid
humours driven by vital processes. From the Inner Canon on, Chinese
physicians, by contrast, composed it mainly of ensembles of functions in the center of the body and a set of circulation tracts throughout. They described these tracts as a system of unspecified form and

ambiguous course that distributed vitality. The branches moved ch'i and other vital fluids (not necessarily liquid) between the limbs, the head, and the central systems, traversing bones and flesh. Authors created other terminology with which they could describe it (but rarely did) as a physical network of conduits.[41]

The central systems controlled metabolic and other spontaneous vital processes. "The subject of discourse, briefly put, is the free travel and inward and outward movement of the divine ch'i. It is not skin, flesh, sinews, and bones."[42] To call this discourse anatomical would thus be misleading. Medical doctrine characterized the systems not as anatomical features but as offices in the central bureaucracy of the body. The point of discourse about these somatic "posts" (kuan) and what they were "in charge of" (chu) was not to describe the incumbents but to specify their duties.

Most, but not all, of the systems were associated with and named after viscera. But the earliest medical writings took the internal organs for granted, and the first classics described them perfunctorily. Later authors saw no need to improve knowledge of them. From the Inner Canon on, organs and tissues figured in medical doctrines as mere correlates of the body's systems of functions, mainly useful in diagnosis and in schemata that aligned parts of the body with physical features of the macrocosm.[43] This emphasis was opposite to that of the post-Hippocratic Greeks. Greek investigations of bodily structures and substances were more systematic and sustained than their inquiries into vital processes. The central concerns of Greek and Chinese medicine made different sorts of questions important.

The Chinese lack of interest in structure was related to the fact that before A.D. 200 physicians did not perform therapeutic surgery, even trepanation. Whether reliance on noninvasive therapy was a cause or an effect we cannot say. Bonesetting and similar procedures remained in the hands of largely illiterate artisans. We know of only one dissection (A.D. 16, by "skilled butchers") for the purpose of investigating anatomy before the eleventh century.[44] Its uniqueness, like the omission of surgery, appears to be due to a taboo against opening the body. Still, one cannot assume that frequent medical dissections would inevitably have deepened anatomical understanding. If directed by questions centered on processes, opening the body more likely would have led to a new range of functional answers, just as Aristotle's quest for formal and final causes affected what his dissections of animals showed him.

What maintains life was not merely the internal circulation of ch'i but a continuous, rhythmic interchange between body and cosmos: "Covered over by heaven, borne up by earth, among the myriad things none is more noble than man. Man is given life by the ch'i of heaven and earth and grows to maturity following the norms of the four seasons."[45] Shared vital rhythms made the body correspond to heaven and earth and kept the three domains in accord. This was a highly dynamic understanding of the three powers.

Because the body interacts with the cosmos, the permeability of its boundaries was an important issue to physicians. The ch'i that fills the universe fills the body as well: "Since ancient times [it has been understood that] penetration by [the ch'i of] heaven is the basis of life, which depends on [the universal ch'i of] yin and yang. The ch'i [of everything] in the midst of heaven and earth and in the six directions, from the nine provinces and nine bodily orifices to the five visceral systems and the twelve joints, is penetrated by the ch'i of heaven."[46] The body may fail by admitting substances that harm it, by keeping out those that it needs, by letting its own vital substances leak out, or by not excreting what it should. Because the ch'i circulation is fundamental not only to the body's growth but to its maintenance, irregularities in it are responsible for pain and disease. Somatic blockages are analogous to failures of circulation in the universe and the state.

The body corresponded to the physical world not only in general but item by item. The Inner Canon elaborated well-established types of correlation. In one dialogue, when the Yellow Emperor inquires of

his minister Po-kao, "I would like to hear how the limbs and joints
of the body correspond to heaven and earth," Po-kao replies, "In the
year there are 365 days; human beings have 365 joints. On the earth
there are high mountains; human beings have shoulders and knees.
On the earth there are deep valleys; human beings have armpits and
hollows in backs of their knees. On the earth there are twelve cardinal watercourses; human beings have twelve cardinal circulation
tracts [and many more details]. In the year there are twelve months;
human beings have their twelve major joints. On the earth there are
seasons when no vegetation grows; some human beings are child-
less. These are the correspondences between human beings and
heaven and earth." The authors are preoccupied not only with things
but with aspects of cyclic change over time.

An interesting passage from the *Inner Canon* is about the correspondence of the two microcosms, **state and body**. It begins as the
Yellow Emperor tells his minister Ch'i-po that he wants to hear "about
the relative authority of the twelve systems of functions associated
with the internal organs, about which is higher in rank and which
lower." A key to medicine, in other words, is the hierarchy of the visceral systems as departments in a civil service. Here is Ch'i-po's reply:

> The cardiac system is the office of the monarch;
> consciousness issues from it. The pulmonary system is the
> office of the minister-mentors; oversight and supervision
> issue from it. The hepatic system is the office of the general;
> planning and strategy issue from it [and so on for twelve
> internal systems].
>
> It will not do for these twelve offices to lose their
> coordination. If the ruler is enlightened, all below him are
> secure. If he nourishes his vital forces in accordance with
> this, he will live long and pass his life without peril. If he
> governs all under heaven in accordance with this, it will be
> greatly prosperous.

If the ruler is unenlightened, the twelve offices will be
endangered; the thoroughfares of circulation will be closed
off and movement will not be free. The body will be greatly
injured. If he nourishes his vital forces in accordance with
this, the result will be calamity. If he governs all under
heaven in accordance with this, he will imperil his
patrimony. Take care! Take care!

This speech systematically describes the body's functions and, at the
same time, reminds the emperor that his ability to rule depends on
his personal cultivation.

The civil service posts are based on how each set of functions fits
in the ensemble of life processes. These posts make up the body's
internal bureaucracy.

The chapter of which this passage is part describes in detail the
functions of the visceral "offices" and their spheres of authority
(while as usual devoting not a word to the physical organs). These
coordinated bureaus are bound together in a way that gives each its
own authority. The somatic civil service directs the functions in the
center of the body and superintends the transportation system that
moves the ch'i throughout.[47]

But why should spontaneous processes need oversight? Ancient
authors did not pose this question; they knew well enough that
bureaucrats always had work to do.

All this differed singularly from the Greek notion that the organs
performed the constituent tasks. Nor was there any Han counterpart
to the Greek preoccupation with a single ruler that governs all
the body's processes.[48] The statement just examined to the effect
that "the cardiac system is the office of the monarch" is about
coordinating a function ("consciousness issues from it"), not a
command. The emphasis in this passage on the sage-king's enlightenment falls squarely in the tradition of discouraging monarchic
activism.

We are dealing with correspondences, not symbols or metaphors in which one thing merely stands for another. Because the state is a little cosmos, Chinese thought of **the cosmos as a state**. Chinese named the earliest constellations after a variety of familiar objects, but in the period that interests us, many asterisms became governmental departments. The polestar even in the time of Confucius was the monarch enthroned in the north. By the Han era, the area around it had become the royal palace. Four other regions of the sky were reorganized as local courts. The panoply of civil service titles they contained formed quite a contrast with the motley collection of heroes and other mythical figures in the European sky.

This is clear enough from the treatise on observational astronomy in *Records of the Grand Scribe*. It is in fact entitled "The Book of Celestial Offices" ("T'ien kuan shu"). It lists the visible asterisms, each a department staffed by stars. The area round the pole star, for instance, is the Central Palace. The list begins with "the constellation of the Celestial Pole, the brightest star of which is the permanent abode of the god Grand Unity [who corresponds to the emperor]. The three stars next to it are the Three Lords [the ruler's paramount advisors], although some identify them as his sons. Curving behind it are four stars. The large star at the end is the Principal Consort; the other three belong to the rear palace [the women's quarters]. The twelve stars that surround all of these, framing and defending them, are the officials who protect the palace. All of these make up the Purple Palace Precinct." This stellar array kept the unity of the uranosphere and the state visible to anyone who looked up at night.[49]

The *Springs and Autumns of Master Lü* brings together **every dimension of resonance**, first in the body, then in the physical environment, and finally in the state:

Human beings have 360 joints, nine body openings, and five yin and six yang systems of function. In the flesh, tightness is desirable; in the blood vessels, free flow is desirable; in the

sinews and bones, solidity is desirable; in the operations
of the heart, mind, and will, harmony is desirable; in the
essential ch'i, regular motion is desirable. When this is
realized, illness has nowhere to abide, and there is nothing
from which pathology can develop. When illness lasts and
pathology develops, it is because the essential ch'i has
become static.

Analogously, water when stagnant becomes foul; a tree
when [the circulation of its ch'i is] stagnant becomes worm-
eaten; grasses when [the circulation of their ch'i is] stagnant
become withered.

States, too, have their stagnations. When the ruler's vital
power does not flow freely [that is, when he is out of touch
with his subjects], and the wishes of his people do not reach
him, this is the stagnation of a state. When the stagnation of
a state abides for a long time, a hundred pathologies arise in
concert, and a myriad catastrophes swarm in. The cruelty of
those above and those below toward each other arises from
this. The reason that the sage-kings valued heroic retainers
and faithful ministers was that they dared to speak directly,
breaking through such stagnations.[50]

This excerpt is about dynamic relationships. Its correspondences
arise from the medical equation of a normal, unhindered circulation
with health and a blocked or static one with pain and susceptibility
to disease. This equation, transferred to the sphere of monarchy,
underscores a persistent theme. Rulers who want their polities to be
as sound as a healthy body and a normally functioning cosmos avoid
blockages by listening to what their best-informed and frankest advisors say, however unwelcome that may be.

It is again clear that Lü's ideal ruler, a self-cultivator and a ritualist, though no activist, was not inert. Untrammeled responsive-

ness—to the cosmos and to his officers and his people—is what sage-hood implied.

The ruler had a double role in this congeries of worlds. He held the macrocosm and the state in concord, but only if he was sagely. Here is a beautiful statement of this paramount role from the *Book of the King of Huai-nan*: "One who is a sage holds in his breast the heart of heaven. His charisma can move all in the realm. His authenticity stimulates all within. His ch'i moves what is in heaven, so that luminous stars appear, yellow dragons manifest themselves, the felicitous phoenix arrives, springs of sweet wine emerge, and propitious ears of grain grow. The rivers do not flood, and the sea does not become tumultuous. . . . But if a ruler opposes the order of heaven and is violent, there are solar and lunar eclipses and irregular motions of the five planets."

He was also a priestly mediator between the cosmos and the spiritual state of his subjects, as the *Springs and Autumns of Master Lü* earlier asserts. This is from a chapter on how the monarch completes the work of heaven and earth, using his officials as his instrument: "The sage, regulating the myriad things, aims to keep [the vitality of his people as endowed by] heaven intact. When vitality is intact, the spirits are harmonious, the eyes, ears, nose, and mouth are sensitive, and the 360 joints are limber. Those who have attained this state . . . are receptive toward every one of the myriad things and encompass all of them. In this they are like heaven and earth. If such a person were as high as the Son of Heaven, that would not make him proud; if he were as low as the commonest fellow, that would not make him unhappy. This is what we call one whose spiritual power is intact."

As a corollary, the ruler, in attaining sagehood and mediating power, becomes a microcosm himself. The king of Huai-nan's eloquence makes this plain: "The sage covers all as heaven does, bears up all as earth does, illuminates as sun and moon do, harmonizes as yin and yang do, transforms as the four seasons do. [In dealing with] the diversity of the myriad things, [in his eyes] nothing is old, nothing new, nothing distant, nothing close. Thus he is able to model himself on heaven."[51]

All these sources reveal how the macrocosm-microcosm relation works, but why does it work? Han readers knew the answer. All the parts of an organic, cyclic universe interact because that is its spontaneous Way. The political and somatic microcosms resonate in harmony with the macrocosm because the ruler mediates the ensemble of spontaneous processes throughout all three.

There we have it: the cosmos, heaven and humanity's surroundings on earth as its parts, the state, the emperor who personifies it, the body that physicians treated and in which philosophers found correspondences, each organism sharing the universal ch'i, all resonating with the others—these ideas took shape during the mighty efforts to end the chaos of the Warring States. They came together in complex synthetic visions as society reached stability in the Han, and they added to that stability. In the Eastern Han the sense of an orderly cosmos fell apart by degrees as the state did. Macrocosmic thought became attenuated as textual annotation largely replaced it, but it was so deeply rooted that it did not die out.

The Concepts of the Sciences

Chapter 2 argued that the government depended on microcosmic conceptions to organize its functions and justify its authority, and state support in turn influenced the emerging doctrines of the sciences. The authors of scientific literature, as it also showed, drew on the literary forms of conventional classics. It is reasonable to expect of the concepts as well that they differ from one field and one period to another, and that patronage or state employment is a component of variation.

There is no obvious order in which to survey the Chinese sciences. There were no fixed relations between them. Thinkers before, during, and for centuries after the Han did not agree on, or even argue about, what those relations should be.[52] That is why we speak of sciences rather than science. This is not a drawback, simply a difference.

Nor was there a pecking order for scientists and physicians. The Chinese case did not resemble that of the Greeks, whose philosophers taught foundational disciplines and ranked all other studies below them. China's technical officials were scattered through the middle and lower levels of a nine-grade bureaucracy; the astrologers, accountants, and physicians had nothing to do with each other. Where a private practitioner of astronomy stood with respect to a physician depended on social status, not on a ranking of technical knowledge.

Prompted by the emperor's religious motives and his striving for immortality, officials of the astronomical bureau first made a complete ephemeris in 7 B.C. or shortly thereafter. For it they computed combinations of cycles—those of the moon and sun that define month and year, of their conjunction in solar eclipses and opposition in lunar eclipses, of planetary motions. These ordered, eternal cycles were those that rolled through the philosophical sources. They proceeded in unison with the seasonal rhythm of the well-led state, abiding year after year.[53]

The computational terminology itself, the very names of the constants and unknowns, reflected the cyclic method. The methods were designed so that minor bureaucrats with little skill in mathematics could carry them out. These step-by-step procedures for making an ephemeris demanded only ability to add, subtract, multiply, and divide. When more elaborate computation was needed, the astronomical treatise gave the operator a table in which to look up the answer, like those in Ptolemy's *Handy Tables*. "Epoch Divisor" and "Day Surplus" are typical terms meant unambiguously to label the roles of quantities in the computation, nonsymbolic equivalents of modern x and y. Such terms were as plain as bureaucratic titles.

In practice, the cycles basic to the calendars turned out to be far from eternal. After a little more than a century, the calculated beginnings of lunar months (that is, meetings of the sun and moon) often were off by a day, and eclipse predictions were failing regularly. Small errors had been adding up as time passed. That was consequential: baleful omens were increasing in number.

As cycle counting failed in this respect, the result was a long series of court proposals for dealing with crises. The last of many committee reports, in A.D. 179 or shortly afterward, took the bull by the horns. It declared that a mathematical analysis of celestial motions was feasible only within limits: "There is no point in rejecting any method that does not conflict with observation, nor in adopting any method whose utility has not been practically demonstrated." This was a call to compromise the metaphysical ideal of eternal cyclic motions and to explore numerical methods not directly based on it.

Astronomers still used cyclic methods where they yielded adequate predictions, but put increased effort into other computational approaches. Successive emperors approved empirical adjustments between macrocosm and microcosm. The outcome was a set of procedures in which the sense of cosmos played a diminishing role.[54] It is no coincidence that this happened as the Han imperial order was falling apart and after the ideal of a single authoritative set of classics had lost its force.

Calculation of celestial phenomena, on the one hand, and their observation and interpretation, on the other, were distinct but complementary (p. 25). The Grand Scribe was in charge of both. Even before the Han period, reports of observations suggested portentous associations, good and bad. To take a Han example, a meteor drew attention to governmental military operations in the provinces. Its southwestward movement indicated that it was advisable to move against a minority people in that part of the realm. In another instance, the constellation Celestial Boat indicated water; a comet moving out of it was a sign of a great flood. The point in both cases was to be aware of the need for action. The actual function of such reports was to begin a discussion among those responsible for making and acting on decisions. Thus omens sometimes reminded an emperor that he needed to consult his advisors and listen to their advice.[55] The modern view of astrology as a specious predictive technology is beside the point, except to the extent that forecasts inferred from portents could, if confirmed after the fact, vindicate decisions to act.

The Grand Scribe's responsibility for mathematical astronomy was an integral part of his duties. Computation moved celestial phenomena from the unpredictable and portentous to the predictable. Once the calendar could routinely time an event in the sky, that phenomenon became part of the regular workings of government, a source of order rather than disorder. In the Western Han era, the new moon visible on the last day of the calendar month, or the old moon visible on the first, was ominous. By the end of the Eastern Han, these phenomena were no longer a warning that the emperor's mandate was threatened (p. 215), so they were no longer reported. This was patently because by that time astronomers were more accurately predicting conjunctions of the moon with the sun, which mark the first day of the lunar month (if the moon sets with the sun, it will not be visible that night).[56]

For Greek astronomers, the reduction of planetary phenomena to combinations of circular motions also amounted to the imposition of order, and occasionally moral implications were drawn from that orderliness. However, in China the meaning of astronomical order was essentially and primarily political. Its moral significance was a corollary of that.

Two dissimilar Han sources for mathematics were the *Mathematical Methods in Nine Chapters*, of the early first century B.C., by and large a textbook of practical problem solving, and the *Gnomon of the Chou*, compiled 50 to 150 years later, which brought together several attempts at simple numerical models of the cosmos (pp. 39,

59). The short writings in the latter are not conceptually elaborate. Of the few explanations it offers for its models, most prominent is the notion that "the square pertains to earth, and the circle pertains to heaven." Each aspect of the model brings into the correspondence not only its own domain but the functions of the try square and compass, which in the hands of a craftsman can impose useful form on raw materials.[57] This circle-square rationale, in which sky and earth are complementary but opposed, was parallel to the one that binds but separates the ruler and his civil servants (see p. 217).

The *Nine Chapters* is a collection of solved problems, with the answer to each and step-by-step instructions for solving it. Its problems are mostly practical in form but sometimes abstract and occasionally fanciful in content. It is one book rather than an assemblage of short ones, but almost certainly a compilation of individual problems into a standard form. The problems are arranged systematically, though not always consistently, some by type of operation (for instance, those involving right triangles or arrays) and others by area of activity (distribution of goods, construction). The book reflects the inverse of the Greek effort to deduce many true propositions from a few axioms. The compilers show with sets of examples that a great variety of problems can be reduced to nine categories, for each of which one computational method will produce solutions. Rather than formally proving that these methods will always work with the pertinent problem type, the compilers make the point with examples, judiciously not claiming that every conceivable problem would fit into the nine.[58]

The book offers algorithms without proofs (which came shortly after the Han).[59] It does not set out explicit concepts or contexts. Only the content and scale of certain problems reveal bureaucratic concerns. Some chapters are unmistakably keyed to the operations of government, such as the construction of city walls and dikes and the movement of tax grain. The first problem in the last category, for instance, involves the taxes of more than forty-four thousand households in four districts.[60] Other groups of problems, such as the division of commodities and the sharing of investments, reflected the small scale of trade by commoners; still others could have occurred to either administrators or businessmen. In part of the book, the view from the center is unmistakable, but this is not true throughout.

The design that marks a textbook is missing from the recently discovered *Book of Arithmetic*, a manuscript of roughly three hundred years earlier. Many of the problems that it contains are similar, reflecting both local administration and private commerce. But this earlier compilation incorporates and roughly categorizes heterogeneous materials without reconciling or standardizing them. It lacks the system of the *Nine Chapters*.[61]

Medicine began connecting with the physical and political realms as Han healers, moving away from earlier religious and occult rationales, used new, explicit concepts to guide their praxis. At the same time, political theorists borrowed from physicians images of metabolism and the *ch'i* circulation. Among other uses, they applied them in writings meant to persuade the emperor to avoid direct involvement in administration.

In the earliest medical writings, especially in recently excavated manuscripts that record practice before the age of explicit doctrine, the language of popular religion, divination, and the occult arts provided the intellectual glue. When physicians later sought rational, secular modes of explaining their therapies, they seldom attacked these rivals, generally ignoring them or relegating them firmly to the premedical past. Secular correspondence doctrines—for instance, arrays of yin-yang correspondences—occur in other manuscripts devoted to physical and sexual self-cultivation as well as medicine between about 200 and 170 B.C.[62]

Doctrines of the high tradition built up a view of health, illness, and therapy quite distinct from that of popular rivals. It is not that the new classical medicine drove out popular therapy; the two continued to coexist throughout history.[63] Physicians did not see themselves as threatened by religious curers (as they did many centuries later). This is no doubt due to the social gulf between them. Classical medicine was not accessible to, much less affordable by, most of the population. It fitted the intellectual horizons of the scholar-officials so well that they formed a most willing clientele.

At the same time, literate physicians drew on and turned their rivals' understandings in new directions. For instance, hsieh originally meant "askew" as opposed to "upright," and it soon came to imply spiritual pollution.[64] After philosophers applied the word to malevolent conduct and it came to mean "heterodox" in state ideology, physicians adopted it as a term for any pathogen, seen as a type of ch'i (p. 197). The first doctrinal writings of medicine imposed the philosophers' complex parallels between cosmos, state, and body on popular traditions of disease and curing that even elite patients grew up with and would find meaningful.

The earliest doctrines of materia medica also drew on political patterns. The *Divine Husbandman's Materia Medica*, probably a little later than the *Inner Canon*, uses a threefold taxonomy of its 365 drugs to merge Han ideals of immortality with those of medicine and politics:

> Drugs of the higher sort, 120 kinds, used as monarchs; for nurturing life; corresponding to heaven [among the three powers, heaven, earth, and man]; without toxic principles. When taken in large amounts or for long periods they are not harmful. Those who want to lighten their bodies, augment their ch'i, and increase their longevity without growing old will use drugs in the higher canon.
>
> Drugs of the middle sort, 120 kinds, used as ministers; for nurturing the nature; corresponding to man; with or without toxic principles, so that one must determine appropriate usage. Those who wish to check medical disorders and

replenish depleted energies will use drugs in the middle canon.

Drugs of the lower sort, 125 kinds, used as assistants and emissaries; for treating medical disorders; corresponding to earth; mostly toxic. They may not be taken for long periods. Those who wish to expel cold or hot pathological ch'i, to break up accumulations [of ch'i in the body], or to cure diseases will use drugs in the lower canon.

The monarchs were principal drugs in compound prescriptions; the ministers were adjuvants; the assistants were collaterals that attack pathological agents, and the emissaries were drugs that guide other ingredients to the sites of certain disordered functions. Lightening the body and augmenting the ch'i halted aging and prolonged life indefinitely. A prescription, according to this schema, included a single monarch, so many ministers, and so many assistants and emissaries. The several drugs in their vertical hierarchy made up a government task force ready to restore somatic order.[65]

This is far from the whole story. Yin-yang, five phases, and other associations finely specified drug action, but there is no need for further detail here to make the point. In medicine, too, the conceptual structure of health, disease, and therapy linked the body to the cosmos and the state.

Alchemy was the science furthest removed from state control and also the one least adequately documented. There is one brief book on alchemy that may have survived from the Han. The *Nine-Cauldron Divine Elixir Canon of the Yellow Emperor*, which claims to be an imperial revelation, is not typical of the classics that chartered other disciplines. Although written in the first century A.D. or later, it probably became a classic only after the Han era (see p. 60).

The author does not appeal to abstract concepts or to microcosmic symbology. He plainly sets out the procedures, ritual and chemical, for preparing seven "divine elixirs." As the adept eats a completed elixir and metamorphoses into an immortal, "he recovers from all illnesses, jade maidens come to serve him, and the Director of Destiny erases his name from the Book of Deaths and inscribes it in the Register of Immortals." He ultimately rises to heaven, accompanied by his whole household. This framework is plainly that of popular religion (with its bureaucracy of gods) rather than of state cosmology or of the theocratic Taoist movements at the end of the Han period.

The instructions, too, are not at all redolent of the palace. Of their two general themes, the first is the need for privacy (unavailable to rulers) and, if possible, reclusion. If one cannot prepare the elixir deep in the mountains or in some other wild place, one must instead rely on high, thick walls. "Be certain to avoid contact with vulgar and stupid people. Do not let the jealous or talkative, or those who do not believe in this Way, learn what you are doing." The second theme is a warning against trusting herbal substances, the standbys of medicine. "Herbal drugs, if you bury them, will rot; if you heat them, they will decompose; if you roast them over a high flame, they will scorch. If they cannot save their own life, how can they give life to humans? They may cure illness and augment ch'i, but they cannot prevent death." This assertion directly counters the Divine Husbandman's view of the highest class of materia medica, which included many kinds of flora.[66]

To sum up, what little we know about the conceptual structure of Han alchemy suggests that its lack of regular state support and its freedom from political constraints are duly reflected in its independence of microcosmic thought.

Conclusions

In this chapter we have examined the relations of Chinese thought about natural philosophy and science to the social and political circumstances described in Chapter 2. We have found them to be interacting parts of a single whole, not items linked as cause and effect. This will become clearer as we summarize the complex answers that have emerged in this chapter to the questions we raised at the beginning of it.

The conventional wisdom in the late Warring States period was a compound of popular religion and aristocratic usages that depended largely on ritual codes. Local rulers did not enforce or exert authority over folk belief, and the customs of the wellborn were breaking down as society moved toward unification and the structure of power ceased to be feudal. There was great scope for new views of order. Although philosophers agreed that there should be an orthodoxy, they did not agree on what its content ought to be. They did not fight with peers in public, but formed their own positions in their coteries and, in general, played out their rivalries quietly.

As they thought about the universe, what intrigued them was its connection to sociopolitical order. Even the earliest cosmologists related the human body and its health to this nexus. By the Han some were paying considerable attention to microcosms and refining an abstract conceptual language to express the relation of the three domains, cosmos, state, and body.

To understand the evolution of cosmology and the emergence of the sciences and medicine as distinct bodies of doctrine, the responses in each author's work to his predecessors and rivals are important, but they are only part of the picture. The communication that did the most to shape cosmology, like other aspects of philosophy from the age of patronage onward, was one-way, meant for a ruler who was not obliged to respond. Intellectual life gradually came to rotate about this monarchic pole. The state's support for philosophers and scientists, first through local patronage and then through the imperial civil service, led intellectuals to use the conception of the state as microcosm even in mathematics, astronomy, medicine, and other technical fields. Thus, in the late Warring States and early Han periods, despite the remarkable diversity of thought, the result was a fairly coherent ordering of experience.

Cosmology and social, political, and ethical thought, as well as the basic doctrines of medicine, overlapped greatly because they embodied parts of a single, sustained intellectual effort. Philosophers since Confucius tried to end what seemed to them endless turmoil. Certain Ch'in and Han philosophers, beginning with Lü Pu-wei before the Ch'in unification, transformed the ideal of the ruler once and for all from a military strongman into a ritualist who through self-cultivation could mediate a dynamic harmony of heaven and earth. The First Emperor of the Ch'in, although he adopted five-phases symbology and revised state rituals to legitimate his victory, relied heavily on coercion.[67] The High Progenitor (r. 206–195), who went from rural sheriff to rebel to founding emperor of the Han, was faced with the problem of how to rule a vast domain once the fighting was over. He accepted his advisors' proposal that ritual is more effective for the purpose than force, although from time to time he also relapsed into coercion and violence. His successors based their polities to a greater extent on ritual. In that sense, all concurred with arguments for institutions built around the monarch's primary role as cosmic mediator—even when they did not actually make that their own primary role.

At the same time, intellectuals drew on all contemporary currents of thought to create new synthetic ideologies that founded the authority of the state on the regularity of the universe. They juxtaposed the order of the state with that of the human body. As they portrayed the ruler, he not only participated in the cosmic order but performed on the body politic the charismatic work of prevention and healing familiar to every one of his subjects who had been treated by a physician.

Gentlemen who wanted to grasp the great issues of their time, to play a part in the quest for order, or simply to attain the high status that came with official appointment, explored not only the good society but the world outside it and the world within the skin. Their agenda for the ruler became maintaining resonance between these domains. Their account of the physical world, their concepts and preoccupations, kept this goal in focus. What became the main problems of mathematical astronomy, for instance, bore on omens and thus on the imperial family's mandate to rule.

Philosophers did not invent technical work. Long before the authors in Lü's entourage wrote about medicine, unheralded healers had created a large range of curative techniques, substances, and rituals—religious, occult, or physical. The analytic categories that shaped later medical doctrines evolved (partly out of these and partly as a reaction against them) through a long process. It began with retainers in local courts of the Warring States (drawing in turn on the beliefs and practices of private diviners). Between the middle of the third and the end of the first century B.C., a succession of philosophers reworked these tools into a small, increasingly integrated set, drawing on therapeutic as well as other kinds of experience. They incorporated assumptions and modes of discourse that were coming to undergird all systematic thought. Although the new conceptions enabled powerful analyses, they aligned them with the state's view of social and political order.

As this exploration deepened, and the division of intellectual labor began and ramified, many practitioners of the sciences, mathematics, and medicine moved away from what by the first century B.C. had become conventional philosophy and into their own more restricted domains of knowledge. Still, these experts had no reason to reject the assumptions and literary forms of their generalist predecessors. The sciences influenced by government support further elaborated the mature conceptual synthesis. Physicians, for instance, used microcosmic conceptions to rationalize a wide range and scale of therapeutic experience. The *Mathematical Methods in Nine Chapters* reflected the ambiguity of mathematicians' relation to the state. The one surviving Han treatise on alchemy, the only unsubsidized science that is documented in the period, does not share these characteristics. All of this suggests that state support and the control that resulted from it strongly influenced intellectual endeavors, just as concepts that favored harmony and symmetry reciprocally shaped the organs of government.

Comparisons make that conclusion a great deal clearer. Chinese philosophy, lacking the competitive abrasiveness that underlay the Greek variety, remained narrower in its range of exploration and more inclined to seek general agreement on basic issues. This makes it easier to understand another striking difference. Hellenic thinkers fundamentally redefined rare words or coined new ones to take the initiative away from their opponents. Elements (*stoicheia*), nature (*phusis*), and substance or reality (*ousia*) are examples (p. 142). Chinese cosmologists instead adapted or combined familiar words to fit new technical contexts, which their older meanings still influenced. Thus they built up a comprehensive account out of materials that were already part of courtly discourse. This was a less disruptive tactic, better suited to an environment in which discretion counted.

A series of Chinese themes superficially similar to the Greek antinomy of appearance and reality actually were formed in very different frames of discussion. These were centered less often on confrontations between equals than on one-way advice to rulers. Such rarefied questions as the character of ultimate reality did not fit in such recommendations. With regard to macrocosm and microcosms, although many of the possible relationships turn up in both cultures, in China the focus of intellectuals on advice to government and the integration of state, cosmos, and body in a single dynamic balance gave the issue a weight that it lacked in Greece.

6 Chinese and Greek Sciences Compared

The strategic questions set at the outset were ambitious ones: Why did China and Greece produce the science they did? What can a study of these two societies tell us about the ways science developed in antiquity? Our investigations have confirmed that the sciences in the two cultures were indeed very different in the six hundred years from around 400 B.C. to A.D. 200. In many cases, the central preoccupations of the inquirers, the way they construed the issues to be investigated, differed. So did the fundamental concepts they used to articulate much of their work. Having now outlined some salient features of the circumstances in which scientists operated, and of the work they produced in those circumstances, what can we say in conclusion about our strategic questions?

We should repeat two caveats before we proceed to some positive conclusions First, it is difficult to generalize across domains and periods. Many individual Chinese and Greek thinkers are too complicated to fit broad patterns. The traits that we propose are widespread, not universal, ones.

Second, our method does not assume that any one-way causal account is possible, from one part of our data to other parts. Many of the thinkers we have studied strongly influenced the climate of opinion of their day, affecting views on what social and political arrangements and ideals are desirable or viable. The interests of the various groups to which they belonged affected them in turn. Nevertheless, the society did not determine the science, nor vice versa. Each culture consisted of a single interactive manifold, comprising politics, society, institutions, practices, knowledge, and theories.

We began our exploration of the Chinese and Greek cultural manifolds in which science developed by examining the social origins and livelihoods of practitioners (Chapters 2 and 3), working our way gradually toward the issues and concepts central in their work (Chapters 4 and 5). In order to summarize our responses to our strategic questions, we will use seven main headings that reveal important similarities and differences between the Chinese and Greek manifolds in which science developed. We will first pick up our discussion in the last two chapters concerning (1) the chief issues that Chinese and Greek investigators tackled and the fundamental concepts they worked out. That relates primarily to the *contents* of their inquiries. We will then turn to summarize some key features of the *aims, styles,* and *milieux* of those inquiries by comparing (2) the prospects of employment, (3) applications of cosmology and science, (4) the possibility of pluralism and attitudes toward deviants, (5) the contrast between the private and the public spheres, (6) the relation between, on the one hand, the desire for consensus and the need for a common language and, on the other, disagreement and dispute, and (7) the management of persuasion. Although there are evident connections and some overlap between these heads, they provide a convenient schema to structure our conclusions

Concepts

According to a widespread view of the history of science, one might expect that the results of competent systematic investigations, even in ancient societies, would be the same or, if not, would rapidly converge. Both Chinese and Greek astronomers studied the skies to regulate the calendar and to determine the periodicities of eclipses and other phenomena; both Chinese and Greek doctors investigated the pulse in their endeavors to diagnose diseases. Yet even those similarities masked divergent assumptions about how the regularities of heavenly motions are to be understood and about the internal functioning of the human body.

The fundamental concepts in play in China and in Greece were strikingly dissimilar. The Greeks focused on nature and on elements, concepts that seem familiar and obvious to those educated in modern science. They invented the concept of nature to serve distinct polemical purposes—to define their sphere of competence as new-style investigators and to underline the superiority of naturalistic views to the traditional beliefs of poets, wise men, and religious leaders. Element theory concerned what ultimately constitutes material objects. Both natural philosophers and medical writers thought they needed to be able to answer that question in order to explain other phenomena that interested them. Yet rational argument alone could not resolve the disputes over elements between atomists and continuum theorists of various kinds.

Chinese investigators had a very different set of fundamental concerns, not nature and the elements, but the *tao*, *ch'i*, yin-yang, and the five phases. Where Greek inquirers strove to make a reputation for themselves as new-style Masters of Truth, most Chinese Possessors of the Way had a very different program, namely, to advise and guide rulers. They, too, had to be more persuasive than their rivals, with means and aims that differed from those of the Greeks. To that end they took over and redefined existing concepts, such as *ch'i*, to produce a synthesis in which heaven, earth, society, and the human body all interacted to form a single resonant universe. A comprehensive understanding of cosmic order undergirded the advisors' insistence on orderly behavior even from rulers. Some rulers accepted, to a greater or smaller extent, the role their counselors cast them in. All accepted the institutions that surrounded them with advisors.

Particularly at the beginning of a dynasty, the monarch's transformation from a wielder of force to the sole mediator between heaven and earth, ritual guarantor of good order in the cosmos, legitimated his rule. The notion that rulers should rely on their advisors went much deeper than rational arguments; images of the body and of celestial order reinforced it. In both instances the concepts that scientists and physicians used were closely linked to what they aimed to accomplish in the world around them.

Livelihood

Philosophers and scientists everywhere and at all times have, no doubt, sought employment and positions where their skills would earn them esteem and compensation. Nor is there any reason to believe that ancient Chinese and Greek thinkers differed on that score in any notable respect. Yet their actual prospects were markedly dissimilar. Chinese rulers accepted the need to listen to or even depend on their advisors. This was already the case with some of the "guests" whom powerful patrons collected at their courts before 221 B.C. Such clients depended on the continuing favor of their lords. Under the empire, intellectuals moved into secure official positions in the imperial civil service.

The opportunities for Greek intellectuals were much thinner. Few held positions of much influence with statesmen, and the state had no appreciable bureaucratic structure until Roman imperial times. Greek philosophers and scientists were only occasionally recipients of patronage. More often, they earned their living by teaching or by practicing such useful skills as those of the doctor, the architect, or the astrologer. Not all Chinese philosophers and scientists were, or

wanted to be, employed by rulers. But for most that was a viable ambition and the surest source of wealth and prestige.

The situation in medicine was typical. In Greece, one city-state or another irregularly employed a physician, but this provided no support for medical innovation. Although Alexandria was a great center of learning in the early third century B.C., there is no evidence that Herophilus and Erasistratus, who carried out vivisection on criminals, were employed by its royal institutions. In China, the advantages of official status encouraged a high level of education in the ambitious. The government, although it did not regulate or publish standards for private medical practice, carefully trained and tested Imperial Physicians.

The difference in sources of livelihood is fundamental. On the one side, the Greeks, with few patrons to please and no official employment, had to fall back on their own resourcefulness in building a reputation and in making a living. They had little access to centers of political power and greater freedom to engage in abstract speculation—and in many cases, little option but to do so. On the other side, the Chinese could hope to influence the affairs of a very large state, but those opportunities mandated a certain circumspection. The roll call of those who fell out with rulers and as a result suffered imprisonment, exile, castration, or death includes not just Lü Pu-wei and Liu An, both key figures in the development of ideas on macrocosm and microcosms, but also the doctor Ch'un-yü I and the imperial scribe, historian, and astrologer Ssu-ma Ch'ien.

Applications of Cosmology and Science

Although Greek thinkers had little political leverage, their cosmologies, like Chinese ones, reflected political ideas and had marked political associations and significance. The cosmologies of the two cultures were equally value-laden and drew on comparisons between the macrocosm and the microcosms of the state and the human body. The Greeks often saw these as analogous; the Chinese, rather, as parts of a complex whole.

Once we probe further, the differences are striking. Chinese generally agreed on the role that the ruler should fulfill. As the mediator between heaven and earth, he was responsible for the welfare of "all under heaven." The cosmos was the state writ large, and the two formed a seamless whole. Greeks, however, did not agree on the best kind of political constitution, nor on what type of political structure the cosmos was or resembled. They, too, used relations between macrocosm and microcosm to advocate their notions of the cosmic dispensation, but as in the political world, so too in the cosmos, ideas of monarchic order competed with oligarchic, democratic, even anarchic ones.

A similar contrast applies to expectations about the public utility of scientists' ideas. Thus when the Greek astronomers accurately determined the lengths of the solar year and the lunar month, the authorities of city-states made only halfhearted attempts to apply their results to calendar making. For the Chinese, by contrast, the regulation of the calendar was a matter of state concern, implicating the emperor's own charisma. The ruler had every incentive to use the best knowledge available. Through the imperial astronomical bureau, he provided institutional support for research on an imposing scale.

Pluralism and Deviance

Whatever the Chinese thought of the virtues or failings of particular rulers, they were united around the political ideal of a wise prince ruling benevolently. A sign of that wisdom was his reliance on loyal and upright ministers. In the chaos that marked the end of the Warring States, the yearning for a stable order was overriding. Union seemed the only prospect for stability. But even under the empire, stability was achieved slowly and did not last long.

In other circumstances, no doubt, Chinese might have invented governmental forms other than centralized rule, although participatory democracy of the Greek type was not feasible in states as large as those that fought for hegemony in the third century B.C. Regardless of what might have been, the pattern remained one of central rule. In a social system that valued civil service above every other career, philosophers who wanted to be politically engaged—or simply respectable—understood the danger of proposing alternatives to the current dispensation of power. Open divergences of view were generally limited to areas that did not threaten the political status quo. Such areas were few and far between, for the government usually treated outsiders with views opposed to any of its policies as disloyal and potentially rebellious.

In Greece, pluralism was not just possible but even mandatory. Aspiring intellectuals, whatever their field of interest, had to make a name for themselves, and aggressive innovation was a way to do it. There was nothing to lose, and everything to gain, by advocating new ideas in every sphere, from politics, through cosmology, to abstract issues such as the philosophy of space or time and the ultimate constitution of physical bodies. The Greeks were accordingly prepared to entertain as many different solutions to the problems of speculative theory as they did to political questions. The debate on whether slavery was or was not a natural institution illustrates both the willingness to countenance alternatives and the fact that some philosophical arguments had no practical outcome.

The pluralism just described did not prevent many Greek writers from advocating the sternest measures to neutralize those they disagreed with, whom they often considered deviants, pathogens infecting the body politic and in need of drastic treatment. Those who took such an authoritarian line (Plato especially) pictured the true statesman as the doctor whose expert advice all were obliged to take—indeed, should be constrained to do so. Such at least was one ideal, though not generally expressed by theorists who were in positions to implement it.

The Chinese were not particularly concerned with deviance in this connection (p. 57). The usual term for pathogen, *hsieh*, referred also to heterodox thought, but the only deviants to whom the early sources applied it as a medical metaphor were ministers guilty of misconduct. Good physicians were, in China, too, often the model for good advisors, but authors applied that analogy in one crucially different way. The Chinese similes consistently stressed the virtue or lineage of practitioners, not their expertise. Anecdotists as well as medical authors took the expertise of exemplary doctors for granted, as in Greece they did not. There, the theories that the doctor should adopt, and the therapeutic procedures he should use, were as much a matter of dispute as any other of the topics on which individuals made their reputations. By contrast, Chinese examples of good doctors tended to stress their embodiment of social norms.

Public and Private Spheres

In Greece, openly favoring one's private interest was acceptable; in China, it was unthinkable. Even the most rapacious Chinese officials mastered altruistic modes of justification. The unbridgeable gulf of power between the emperor and his civil servants encouraged another level of mendacity. Officials were tools of imperial control, on the one hand, and, on the other, representatives of a hereditary elite with certain interests, such as adding to the prosperity of one's clan, that were inherently opposed to those of the ruler. Even if the historical record does not openly acknowledge this tension, it often inspired peculation and worse.

The sphere of operation for most Greek philosophers and scientists of every kind was the private, not the public—reflecting the point already made that the possibilities of working in the state domain were so much more restricted. Greek architect-engineers could hope for state commissions, and there was a little scope for public doctors. Yet such positions carried neither security nor any assurance of a successful career. Reputation depended on repeated successful self-presentation far more than on the personal favor of a ruler or his ministers. In the classical period in Greece, the authority responsible for awarding architectural commissions and for the appointment of public doctors was the assembly of citizens as a whole or the elected council acting on their behalf. Aspiring architects and doctors knew they had to persuade their fellow citizens. Even when that situation changed, with the rise of the Hellenistic kingdoms and then Roman domination, the appointees never enjoyed stable employment as officials in an elite bureaucracy.

Consensus and Disagreement

Ancient Greek culture encouraged disagreement and disputation in natural philosophy and science as in every other field; the Chinese emphasized consensus. This was not because the Greeks were intrinsically disputatious and the Chinese essentially irenic. Success in debate was how you made your name in Greece, in a way that has no analogue in China. A few Chinese thinkers before 250 B.C. were consistently critical of others, and many throughout our period carped from time to time. Conversely, Greek minds met on fundamentals often enough to make some aspects of their early thought cumulative, in mathematics and the exact sciences especially. But even when, in the Hellenistic period, more stable groupings of philosophers and of doctors formed, they were locked in constant competitive debate with one another. In medicine especially, even when Greeks agreed on certain ideas and practices, they often gave them different rationales or justifications.

In China before A.D. 200, general agreement on a broad range of issues was more common, with Chuang-tzu the only consistent clear exception. Everyone else favored orthodoxy, although philosophers had very diverse notions of what its content should be. A consensus on broad principles did not imply that doctrines were standardized. Within the medical terminology based on yin-yang, the five phases, and analogous concepts, there was much room to vary elaboration on points of detail. At that level, Chinese did not strive for complete uniformity. What would have struck Greeks visiting China, however, was that widely shared technical languages with generally agreed meanings provided a framework within which scientists could develop ideas.

From the Han period on, followers of the classicist traditions seldom rejected old systems as they proposed new ones. Scholars of medicine and other fields were more intent on reconciling discrepant systems of correlations than choosing once and for all between them. As they added to or commented on earlier ones, they produced increasingly comprehensive, complex schemata. There was no single occupational or educational structure to enforce one authoritative synthesis, so general agreement on the outlines of doctrine never extended to fine details.

Nor can one assume that physicians wanted total agreement. Multiple explanations of the same phenomenon, often inconsistent ones, remained frequent. Although a logician would condemn the resulting doctrines as unrigorous, Ch'i-po defended them as a fuller account, more adequate to the complexities of medical practice than narrow consistency could have been. Chinese preferred on the whole to cascade levels of meaning and build the richest feasible explanations rather than seek a single cause that ruled out all others. For that reason, doctrinal divergences seldom inspired debate.

The levels of overt disagreement therefore differed markedly in China and Greece. Where both ancient societies differ from the modern situation is that in neither were there—nor could there have been—professional institutions of the modern type, set up to settle technical difficulties, standardize language, oversee qualifications, and, in medicine especially, control access to the profession and the practice of its licensed members.

Persuasion

There is, finally, the question of the uses of rhetoric, where issues of status are always in evidence.

For Greeks, whatever other purposes oral contention served, it was a tool of competition. Lacking sinecures or even secure employment, Masters of Truth depended on debate for fame and livelihood. Argument tended to be face to face, and the public was often expected to decide, just as the public determined the outcome of political discussion in the Assemblies and of litigation in the Dicasteries. These usages had no Chinese counterparts.

In China, Possessors of the Way, with some exceptions, hoped for support from rulers, as clients in the late Chou courts and as officials in the Han. Their interlocutors were not fellow thinkers but their patrons and employers, who expected advice but did not feel obliged to act on it, nor even to reply to it. This relationship hardly made for lively exchanges, and few were recorded. Open disagreements with fellow philosophers and scientists (except on matters of state policy or heterodoxy) were negligible by comparison with Greek practice. Most philosophical disagreements were written, and were directed at dead or absent rivals.

The lineages of China avoided internal disagreement except in affirmatively developing the ideas of intellectual ancestors. Disciples, with a very few exceptions in the Warring States era, did not openly reject their masters' teachings. Chinese coteries, unlike Greek philosophical and medical sects, avoided publicly declaring war on each other. They preferred to imply their own superiority by conceding that others had a partial—but only a partial—grasp of the Way (e.g., p. 54). Hostilities were unproductive when teachers (not to mention parents) aspired above all to public employment for their pupils. Neither celebrity nor the esteem of all colleagues everywhere could compensate for lack of access to appointments. In China, teaching ranked well below official service on the scale of prestige; scholars generally accepted it as a livelihood out of economic necessity. In the first and second centuries A.D., official registration of teachers gave a great advantage to those who already had political status. In Greece, pedagogy was an essential part of the activities of philosophers and scientists of all types.

The Way and the Word

How individuals in China and in Greece responded to the complex traditions they inherited, how they made and used their own room for maneuver, negotiated their own positions, decided their goals and set about fulfilling them: these are not questions to which a single answer is possible. Rather, both societies offered a wide spectrum of possibilities for aspirants to the *tao* or to *logos*.

Neither China nor Greece had a monopoly of the wherewithal to develop science. Both had ample conceptual tools and institutional frameworks to engage in systematic inquiries into the sky, the human body, the cosmic dispensation as a whole. Each exhibited its own distinctive potential for the pursuit of such investigations. The dominant, but not the only, Greek way was through the search for foundations, the demand for demonstration, for incontrovertibility. Its great strengths lay in the ideals of clarity and deductive rigor. Its corresponding weaknesses were a zest for disagreement that inhibited even the beginnings of a consensus, and a habit of casting doubt on every preconception. The principal (though not the sole) Chinese approach was to find and explore correspondences, resonances, interconnections. Such an approach favored the formation of syntheses unifying widely divergent fields of inquiry. Conversely, it inspired a reluctance to confront established positions with radical alternatives.

Our hypotheses will need further testing in relation to these two societies and, further afield, to other ancient and early modern cultures. Despite its limitations, this collaborative exploration of Chinese and Greek science, we believe, has identified some fundamental factors that can help to tell us why these societies produced the science they did.

Appendix

EVOLUTION OF THE CHINESE COSMOLOGICAL SYNTHESIS

A basic framework for discussing cosmology developed in three stages. In the first, officials and clients in local courts, beginning in the eighth century B.C., invented and adapted a wide variety of categories to talk about ritual, the cosmos, and exigencies of state policy. In the second, third-century intellectuals, defining a central monarchic order, narrowed their focus to a small number of concepts and explored how they might fruitfully relate them. Authors derived their concepts to a large extent from those of the first stage, understanding them in a more technical fashion, but also drew on themes that had appeared in commentaries to the Book of Changes. During the third stage, in the first century B.C., two remarkable books elaborated the idea of ch'i as a universal principle of action as well as a universal substance and used yin-yang and the five phases systematically to analyze processes involving ch'i. This became the norm in philosophy, and general in writing on science and medicine.

Fundamental Concepts: The First Stage

Reconstructing the thought world of the Warring States (408–221 B.C.) is a difficult business, particularly at a time when scholars are

reevaluating the dates of almost all early sources. The most important is *Master Tso's Tradition of Interpretation of the Spring and Autumn Annals*, a
great collection of reports and anecdotes dated between 722 and 464
that elaborate on the laconic records of the *Annals*. Persons unknown
compiled it, in all likelihood around 310, incorporating early
materials. The editors reworked those materials, to what extent we
do not know. The text reflects only to a very limited extent the great
changes in syntax and diction between the eighth and fourth centuries. Our working hypothesis is that the editors kept the content
and frames of discussion fairly intact while homogenizing the language. This book, together with two other traditions, those of a
Master Kung-yang and a Master Ku-liang, can be used cautiously to
throw light on the differences between Warring States concepts and
those of the third and second centuries B.C., the second stage.[1]

In the conversations between astrologers, diviners, military
commanders, physicians, nobles, and rulers in the *Tso Tradition*, we
overhear novel explanations of both social and physical phenomena
using numerical categories. For some time such categories had
been implicit in court ritual, both in its structure (orientation on the
four cardinal points, and so forth) and its performance (gifts to
humans and offerings to divinities presented in set numbers).[2] As
early as the eighth century B.C., courtiers extended numbered rubrics
to every sort of subject matter. In particular, when officials were
broaching ideas to the ruler or suggesting ways to justify policy, this
numerology cast new themes into a form that ritual had already made
familiar.

An account bearing the date 710 in the *Tso Tradition* illustrates this
extension of language. A retainer is chiding a lord who has wrongly
installed a bronze vessel in the Chou king's ancestral temple, a grossly
improper act. He points out that the authority of a ruler depends on
maintaining propriety in everything, down to the hues of his robes,
"the five colors of which reflect counterparts" in the external world
"in order to illuminate their association with things in it."

A discussion in the Ku-liang Tradition of abnormal weather suggests an early stage in the cosmic notion of yin and yang. These two words originated as concrete terms for the sunny and shady sides of mountains and streams, a sense that recurs still in Han writings. The *Annals* records a great thunderstorm in the late winter of 714 B.C. and a heavy snowstorm a week later. The *Tradition* comments that having "two major divergences from the norm in eight days is due to the intertwined motion of yin and yang." What this means is uncertain, but it clearly does not mean the interpenetration of sun and shade.[3] Joining the two as complementary opposites began the long process that metamorphosed yin and yang into the abstract yin-yang (pp. 198–99).

In other passages from the Tso Tradition, groups of twos and fives, not at all standardized, appear in profusion. So do groups of threes, fours, sixes, sevens, eights, and nines, as in an example that refers to events in 517. A high minister explains that the three powers (p. 215) provide a cosmic basis for ritual:

Ritual is founded on the regularities of heaven, on what
is appropriate on earth, and on the actions of humans.
Humans take the regularities of heaven and earth as
their pattern. . . . Heaven and earth give rise to the six
ch'i and make use of the five agents (wu-hsing). The ch'i
give rise to the five flavors and emit the five colors, and are
manifested in the five modes of music. When the [responses
of humans] are excessive, they lead to confusion, and people
lose their proper natures.

For this reason, the sages created ritual to uphold proper
distinctions. They designated the six domestic animals, the
five animals that are hunted, and the three animals that
are sacrificed in order to uphold [the distinctions between]
the five flavors [of the offerings]. They made the nine
ornamental designs for robes, their six hues, and their five

types of display to uphold [the distinctions between] the five colors. They made the nine songs, the eight popular songs, the seven sounds, and the six standard tones to uphold [the distinctions between] the five modes.

This report leaves the impression that the denizens of the local palaces were already basing ritual on the harmonies of the external world. They were fashioning a social order over which a ruling family presided and "reinforced the natures of heaven and earth so that they would endure for a long time."[4] An order founded on abiding cosmic harmonies—for that is what the numerical categories embodied—would last, whereas those that depended on force or arbitrary acts would perish. The point of these diverse categories was, in other words, an elaborate ritual framework that would ensure the survival of a polity.

Rituals had employed sets of regalia and objects since the second millennium B.C., but we have no reason to believe that the rites of the early Chou dynasty were based on correlations like these. This macrocosmic justification explicitly based on arrays of numbers was strikingly new.

The same main categories appear in a text that may also reflect sixth-century usage. A physician is diagnosing the illness of a ruler as due to sexual hyperactivity. The doctor reveals the connection between disease and the environment:

In the external world there are the six ch'i. In descending they give rise to the five flavors. They emit the five colors and are manifested in the five modes of music. In excess, they produce the six illnesses.

These six ch'i are shade (yin) and sunshine (yang), wind and rain, dark and light. In their divisions they form the four seasons; in their sequence they form the five nodal points of the round of the year. Their excess is responsible for calamities. Excessive shade is associated with cold

illnesses, excessive sunshine with hot illnesses, excessive
wind with disorders of the extremities, excessive rain with
disorders in the belly, excessive dark with disorders involving
mental confusion, and excessive light with disorders of the
heart and mind. Women are things of sunshine, but darkness
is the proper time [for sexual intercourse]. When it becomes
lewd, that gives rise to internal hot disorders, mental
confusion, and demonic infestations.

In this instance, my lord has been neither restrained nor
timely; how could your conduct not have led to this?[5]

In the first stage of cosmology, there were many fivefold groups,
often inconsistent, among the numerical categories that court
diviners and others were trying out. The fives tend to be as concrete
as yin and yang, still largely shade and sunlight. They are rooted
in the flavors of food offerings, the colors of ceremonial robes, the
modes of ritual music. The courtiers ordinarily speak of "five
materials" (wu-ts'ai) and only twice, in apparently irrelevant contexts,
use wu-hsing, the word that later came to mean "five phases."[6] A complicated process led from these notions to the cosmological five
phases.

"Five materials" occurs in the Tso Tradition in anecdotes relating to
the years 546 and 531 B.C., which may mark the early beginnings.
In one example the political point is clear: "Heaven engenders the
five materials, and the people use them all. They could not do
without any one of them. Who could do without weapons? Weapons
have been provided for a very long time, in order to awe those who
do not follow the established ways and to show off cultivated virtue."
Some Han commentators, in hindsight, identify these five as the
substances—wood, fire, earth, metal, and water—on which later
writings based the notion of five phases. Most modern scholars
follow this reading. Nevertheless, the most erudite Han scholiast,
Cheng Hsuan, when explaining this passage, enumerates the five

more plausibly as the actual constituents of weapons: metal, wood, leather, jadelike minerals (yü), and clay.[7] It is impossible to be sure what the term meant in the sixth century. Scholars' arguments from the Han era to the present day have not resolved the issue.

An important point of comparison is clear. Nothing in discussions of this term resembles Aristotle's theory that every piece of wood or bone (or anything else) contains earth, water, air, and fire as its elements. Nor are these courtiers speculating along the lines of the indivisible atoms of Leucippus and Democritus. The "five materials," whichever five they may be, are just materials.

There is another politically important set of fivefold categories. In the surviving fragments of Tsou Yen's (p. 264) writings, dating to around 300 B.C., the author uses the conception of five powers (wu-te) as an important part of his theory of history. The sources consistently identify them as the familiar wood, fire, earth, metal, and water, but Tsou clearly was not thinking of them as materials or functions that affect materials. He wrote of them as emblematic kinds of activity. They marked new beginnings within a cycle of dynastic succession, determining what colors of court robes and other appurtenances of court ritual each dynasty should adopt. He was, his biography tells us, adapting these emblems to the reform of politics and state rituals.[8]

Lü Pu-wei adapted Tsou's doctrine to new circumstances (p. 263). Lü's lord, the First Emperor, upon uniting China in 221 B.C., adopted Tsou's schema of five powers to regulate rituals and institutions by identifying his new regime symbolically with the power Water. He thus aligned it with the cycle of dynasties that symbolized succession by conquest.[9] The early Han court continued this Ch'in usage. The term "five powers" was still current around 100 B.C. during intensive discussions of how to change governmental practices (see p. 66). None of these uses pursued Tsou's goal of political reform. Their aim, rather, was to strengthen the authority of the state.

The Chinese ensemble of five phases, used, like yin-yang, to analyze processes and configurations, had no Greek counterpart. The conception, though occasionally mistranslated "five elements," had nothing to do with elements. It originated in the sphere of morality and became part of cosmological thought only around 240, the time of Lü Pu-wei.

Hsun-tzu, in the middle of the third century B.C., attacked two dead predecessors, Confucius's grandson Tzu-ssu and Mencius, because "they drew on antiquity to fabricate a doctrine that they call wu-hsing" and claimed falsely that Confucius originated this pernicious teaching.[10] No such doctrine has survived in writings of the two.

A group of short texts from about 280 B.C. or earlier, lost but excavated in two versions in 1973 and 1993, solves this mystery. It speaks of five hsing—benevolence, righteousness, propriety, judgment, and sagacity—and the innate feelings that lead to them if nurtured. These turn out, in fact, to be a minor modification of Mencius's famous "four beginnings" (ssu-tuan). One may conclude that two centuries after Confucius, in certain circles, wu-hsing was a set of ethical categories.[11]

Given the uncertainty of current datings, we can only suggest tentatively that "wu-hsing" first appears in cosmology in the "Great Plan" (p. 215), perhaps a little later (or conceivably earlier) than the Tso Tradition (ca. 312 B.C.).[12] This speech of advice to a king reiterates heaven's revelation to the legendary emperor Yü, who in archaic times saved his realm from the primeval flood. The "Great Plan" offers a panoply of fivefold and other numerological correspondences that form a basis for a new polity. These include the wu-hsing. Although called water, fire, wood, metal, and earth, they are not substances but rather a set of characteristic functions: "Water means soaking downward; Fire means flaming upward; Wood means bending and straightening; Metal means conforming and changing;

Earth means accepting seed and giving crops. Soaking downward creates the salty; flaming upward creates the bitter; bending and straightening creates the sour; conforming and changing creates the pungent; giving crops creates the sweet." Although this association with the five flavors seems to be the start of a chain of correspondences, the speech moves on to other fivefold but unconnected matters, all of them relevant (so the speaker claims) to keeping order in the realm.

Although, at this early stage, "water" and the others (except fire) are the names of materials in other contexts, this author unambiguously defines them as functions. These five, each standing for a characteristic activity, seem to prefigure the Han understanding of the five phases. In this document, however, the "five activities" are not part of a single process. The conception still belongs to the first, experimental stage of development.[13]

Our Warring States sources have revealed a world preoccupied with the dilemmas of political practice, a world teeming with conflict but seeking a harmony expressed in cosmic metaphors. The protagonists, including officials and clients of many sorts, freely tried out concepts of great generality, building diverse numbered categories of officials, activities, and phenomena.

We can speculate about the social dynamics of this innovation. The anecdotes in the Tso Tradition, the "Great Plan," and allied sources, which deal with a time before the efflorescence of patronage, put the new categories in the speeches of officials at the feudal courts. Whether prime ministers insisting on strictness in ritual or lowly physicians arguing for moderation in sexual conduct, all were forming language that would make their advice persuasive. In an anecdote related to the year 488, a group of court astrologers are running through a repertory of cosmological correspondences, complex wordplay, and the esoteric analysis of names to argue for a shared interpretation, only to be trumped by a high official using the Book of Changes to justify his own.[14] New kinds of argument, if they could not always outweigh technical functionaries' lack of hierarchic power, were sometimes effective tools. Some high officials, too, understood their use in persuading rulers who might be annoyed by appeals to virtue.

This competition for influence in individual courts was going on in many states trying to survive amid constant hostilities. The officials of these states were aware of what was happening elsewhere, and a wide array of new correlations appear in the ongoing conversations in all the courts. The spokesmen who invented the vocabulary of Warring States cosmic discourse had no reason to work out a single set of systematic concepts.

Convergence: The Second Stage

By the third century B.C., there were a few intellectuals among the diverse throng of "guests" (p. 28). To form their own abstract schemata, they began to adapt certain notions from the great variety circulating in the regional courts. The concepts on which they soon focused were ch'i, yin-yang, and the five phases.

These concepts became central to the political microcosm that led to the Han syntheses. For simplicity, we will consider only three exemplary authors, Lü Pu-wei, Liu An, and Tung Chung-shu, introduced in Chapter 2. These three, who were active between roughly 240 and 120 B.C., stand out among those who drew on the ideas of Confucius, Lao-tzu, and many others to invent new doctrines of the state.

Lü was a powerful official, Tung a middling one, and Liu a king, uncle of the emperor. Each of the three sought in his own way to remake his monarch into a quiescent figure at the common focus of the cosmic and social realms. It was not that emperors were autocrats. They could not ignore the unanimous decisions of the highest officials or the interests of the most powerful factions.[15] On the other hand, the arbitrariness to which rulers were entitled could make stable governance impossible, and civil servants below the top ranks had no institutional voice.

These three had the bad fortune to write for two of the most irre-
pressible monarchs in early China. The First Emperor, shortly before
his state of Ch'in unified China, drove Lü to suicide; the Martial
Emperor of the Han did the same to Liu (p. 33). The same monarch
enacted some of Tung's proposals (e.g., p. 266) but treated him con-
temptuously and had no interest in embodying his model of the ruler
as Confucian self-cultivator. Despite the personal disasters of the
three, their doctrines of the microcosmic state came, over time, to
form the abiding basis of Chinese monarchic thought.

Lü compiled his *Springs and Autumns* at a time when there was no
longer a Chou king and no one could be sure when the wars would
end. He and his clients distilled from all the political theories of their
time a set of proposals for what would come after the last battle.
They had seen military force establish polities and fail to maintain
them. Their new doctrines transformed the ideal ruler from a con-
queror into a provider of life to his people, guarantor, through his
spiritual cultivation, of a ritual order that would not self-destruct.
The political order became durable because it was not arbitrary but
based on heaven, the eternal standard. Dissidence or disobedience,
contrary to that standard, was bound to fail.

The most prominent feature of the *Springs and Autumns of Master Lü* is
a series of twelve chapters, each one of which sets out for one month
a rich array of mythological, cosmic, terrestrial, and ritual correspondences: "In the first month of spring, the sun is in the lunar
lodge Chamber. At dusk the lodge Triad crosses the meridian; at
dawn, Tail does so," and so on. We need not dwell on the details to
see the general pattern. The place of the sun on its yearly orbit, the
periodic movement of the stars that mark the twenty-eight nightly
lodges of the moon, the days of the ten-day week, the various fivefold arrays from legendary emperors to flavors, their members pre-
dominant one by one as the cycle plays out: all mark a stage in the
ritual round of the year.

Details follow of what the ruler is and is not to do in carrying out the annual round of rituals that keeps the activities of the realm in concord with those of the universe. Thus in the wintry first month, killing and war are to be suspended, no animals are to be slaughtered, no crowds gathered, and no city walls built, all out of respect for the yin character of the season. These cyclic rules that govern heaven, earth, and man were neither new nor unique. Lü did not legislate them in Ch'in. He merely asserted that the state that adopted them would achieve legitimacy and success. Some regimes of the Han and later followed them more or less; many did not. But the ritualists of each era accepted the general aim of achieving the proper cosmic orientation and adapted it to their own means.

Lü applies the same master concepts to medicine: "Heaven engendered yin and yang, cold and heat, dry and wet, the transformations of the four seasons, the metamorphoses of the myriad things. Every one of these can bring benefit, and every one can bring harm. The sage investigates what is favorable in yin-yang and what is beneficial among the myriad things for the good of human life. . . . For nurturing vitality nothing is as important as understanding these fundamentals. Once one understands them, there is no way in which disease can enter."

Lü significantly brings the mutual conquest sequence of the five phases to bear on the ritual changes that accompany a change of dynasty: "When a kingly house is about to rise, heaven invariably manifests auspicious signs in advance to the people below. In the time of the Yellow Emperor, heaven first made large earthworms and mole crickets [*Gryllotalpa* spp.] appear. The Yellow Emperor said, 'The Earth ch'i has conquered.' Because the Earth ch'i had conquered, among colors he honored yellow, and among activities, those related to the phase Earth." An analogous change of phase presaged each dynasty, Lü tells us, and its sage rulers responded accordingly. Thus the governing phase of the Chou dynasty became Fire.

When China is finally unified, "what will take the place of Fire is bound to be Water [since water puts out fire]. Heaven will first manifest conquest by the *ch'i* of Water. When the Water *ch'i* has conquered, among colors [the ruier] will honor black, and among activities, take those [related to] water as his standard." This is exactly the change in court robes and other ritual matters that Lü's state of Ch'in adopted when, after his lifetime, its ruler established the first Chinese empire.

In the passage just quoted—and nowhere else in the book—the five phases are no longer merely a set of five categories that can be applied to a broad range of dynamic phenomena. Lü has made them characteristics of *ch'i*. Ch'i is what changes. Earth, Wood, Fire, Metal, and Water have become abstract phases in the cycle of change. Earth *ch'i* is not likely to be earthy. It is rather the balance point in any dynamic process and may subsume an infinite number of balanced phenomena. He does the same thing with yin and yang, as when he asserts in the same passage that "if in the second month of spring . . . [the ruler] promulgates an order appropriate to winter, the yang *ch'i* will cease to predominate. The grain will not ripen, and his people will often rob each other."

Lü thus related *ch'i*, yin-yang, and the five phases, but this one occurrence in a large book by multiple authors was not enough to define the association. Like Tsou Yen's older (but lost) philosophy of history, the passage just quoted is concerned with political ritual rather than with the physical world.[16]

The *Book of the King of Huai-nan*, a century later, applies yin-yang and the five phases to order broader ranges of phenomena than its predecessors had done. It links yin-yang and *ch'i* fairly often, but only once, in passing, does it associate the five phases with *ch'i*. It is innovative in other ways—for instance, in systematically digesting the astronomical and topographic knowledge of the time. But even as it argues for the new cosmic model of rulership, it speaks of the old hegemonic rulers, who had ceased to exist before the Han era, as if they still existed, and ignores other social changes that a century had wrought.[17]

Most of Tung Chung-shu's *Abundant Dew on the Spring and Autumn Annals,* in its extant form, was written by different hands at different times. The less doubtful chapters identify yin and yang as ch'i no more systematically than Lü's and Liu's books do, and do not link the five phases to ch'i at all. Several chapters that discuss the five phases but not yin-yang do clearly assert a linkage of the former with ch'i. But it is doubtful that any of those were written before the early first century A.D., and some could be later than that. A passage in a five-phases chapter offers one of the clearest formulations of the third stage of synthesis, rather than the second stage typical of Tung's own writings: "The ch'i of heaven and earth combine to form a unity. Divided, they become yin and yang; subdivided, the four seasons; set out in order, the five phases. Hsing means activity. Their activities differ, so they are called the 'five hsing.' The five phases are [equivalent to] five official posts. If one takes them in sequence, they engender each other; taken alternately, they overcome each other. That is why one speaks of them as ordered. If one contravenes them, the result is disorder; if one conforms to them, the result is order."[18]

The books of Lü and Liu and the few defensibly authentic writings of Tung are typical of the many works, less preoccupied with cosmology, that survive from between 250 and 100 B.C. They portray a world governed by the associations of individual phenomena and things with ch'i, the five phases (except in the case of Tung), or yin-yang. They do not fully integrate these three concepts, and their correspondences are often inconsistent.[19] Still, their focus on three concepts out of the many that their forerunners had tried out, their abstractness, and their linkage of ch'i with yin-yang were important beginnings. The analytic language they shaped was equally applicable to the cosmos, the state, and the body.

Their arguments soon proved useful to statesmen, as an example makes clear. First, Tung Chung-shu argued in a series of memorials against continuing the Ch'in dynasty's reliance on the five-phases conquest cycle to determine the ritual basis of government. This usage in earlier Han reigns, he asserted, implied that the dynasty's claim to legitimacy, like that of its detested predecessor, was nothing more elevated than conquest. The continuity also implied that Ch'in was the Chou's legitimate predecessor, a notion that Tung strongly refuted. He created a new Triple Concordance cycle, analogous to the generation cycle of the five phases but with only three phases, to express the freshness of the Han's mandate. He pointedly omitted the Ch'in, implying that heaven had shifted its mandate directly from Chou to Han. In 104 B.C. the Martial Emperor enacted a compromise between this idea and the established five-phases conquest cycle, combining Tung's key proposal of a new calendrical system with a new set of ritual associations taken from the five-phases cycle.

This concession ignored Tung's larger point: because what mattered in the transfer of the mandate was not force but the virtue of the ruling house, only a cycle of generation and nurturance, not one of conquest, could represent it. But the discussion of this point went on for another century, ending in a second change of aegis that revived the Han period—based not on Tung's own idiosyncratic cycle but on the generation order of the five phases.[20]

The Book of Changes as a Map of the Cosmos

A distinct set of contributions to the mature cosmological synthesis came from commentaries to the Book of Changes. The Changes is the oldest Chinese book that has been passed down to the present. Its unknown compilers brought together, a little before 800 B.C., fragments of a divination manual with a jumble of oracles, proverbs, and snatches of rhyme and song. As one would expect in a book so ancient, there is not an abstract concept in it.

The quest of the ju for archaic authority led them, probably in the mid-fourth century B.C., to begin claiming classical status for this work. From about 300 on, they and others proceeded to interpret—

that is, read philosophy into—the nondescript text. Its antiquity implied sagely wisdom that could guide conscientious action, wisdom that, since it was not obvious in the jumbled parts, must be arcane.[21]

At first, scholars concentrated their analyses on the sixty-four main texts (the "judgments"), the main outcomes of divination; each judgment was attached to a six-line diagram (a "hexagram"). They also paid attention to the six secondary texts attached to the individual lines of a hexagram. Each of the six lines could have two values, broken or solid, which scholars inevitably associated with yin and yang. This is a typical hexagram:

They saw the structure as dynamic, because divining with yarrow stalks to choose a hexagram involved building up the six binary lines in sequence. Between the third century B.C. and the second century A.D., scholars set down their interpretations in a group of treatises later called the Ten Wings that were transmitted with the archaic text, and in other writings unknown in later times until they were recently excavated.

Students of the *Changes* found meaning not only in the judgment texts and the six-line diagrams attached to them but also in the eight possible three-line diagrams ("trigrams") that resulted when the hexagrams were split in half (the two trigrams in the diagram above are mirror images). The archaic sages, most came to believe, had originally constructed the sixty-four hexagrams by pairing trigrams. The commentators then applied a rich set of binary (yin-yang), three-valued, six-valued, and eight-valued meanings to the text *and* symbols to add new layers of political and cosmic meaning.[22]

The "Appended Statements," an interpretation that makes up two of the Ten Wings, describes the Changes as the first step of a legendary monarch toward inventing culture: "In antiquity, this is how Fu Hsi ruled his realm. Looking up, he contemplated the images in the heavens; looking down, he contemplated the images on earth. He contemplated the markings of birds and animals and the characteristics of the earth. Near at hand, taking them from his body, and at a distance, taking them from other creatures, he first created the eight trigrams to attain the power of the gods and to arrange in order the actualities of the myriad phenomena. He invented the knotting of cords, making nets to enable hunting and fishing . . ." There follow in sequence Fu Hsi's inventions of agriculture, commerce, and all the other appurtenances of civilized life. Thus the commentary affirms once again that culture (and technology as part of it) is something rulers create and grant to their subjects. Just as remarkably, it sets forth the claim that writing came from Fu Hsi's observation of signs in the material world rather than being derived from speech.[23]

Two experts of the first century B.C., Meng Hsi and Ching Fang (see Table 2 in Chapter 2), pushed further the quest for deep metaphysical consistencies and regularities. They looked for an order hidden in the hexagrams, their two trigrams, and their six lines that in some subtle way determined the words of the judgments. That is, the iconic hexagrams and their constituents became the primary symbols, and the verbal judgments simply conveyed (or hinted at) the truths that the structures of the diagrams express. The Han scholars of the Changes were finding in this archaic book the regularities that govern experience of the external world and everything else. By the end of the Han, their accumulated reinterpretations had metamorphosed the Book of Changes into a universal repository of cosmic thought and monarchic theory and made it influential in the trend toward synthesis. Their approach to interpretation remained a staple of cosmological discussion for two millennia.[24]

The Mature Synthesis: The Third Stage

The exploratory alignments of yin-yang, the five phases, and ch'i in the second stage, as well as new approaches to interpreting the *Changes*, bore fruit in two remarkable books. Yang Hsiung, a court poet and classicist, turned the prevalent cosmic concepts in a more subtle direction. His long prose poem, the *Supreme Mystery*, completed in 4 B.C., was not only deeply integrated in a new way but sensitive to the dysfunctions of the state in his time. The anonymous *Inner Canon of the Yellow Emperor* also elaborated a view of the universe in resonance with the microcosm of the body and with state ideology. Scholars are still debating the date of this book; it may be as much as a century earlier or later than the *Supreme Mystery*.

The *Supreme Mystery* was a realization of the book that scholars had long been reading into that medley of pre-philosophical fragments, the *Book of Changes*. The Ten Wings had reenvisioned the old classic as arcane wisdom without altering its literal disorder. What Yang did was to create it anew, this time structuring it meticulously. His prose poem traces, in exquisitely graded and perfectly regular sequence, the dynamic cycle of the year, divided into eighty-one equal parts of roughly four and a half days each. Each of the book's sections, or heads, begins with a head text in which Yang discusses the activity of yin or yang ch'i in those four-odd days, explicitly correlating them with the five phases. A run of three head texts, which chronicle the first stirrings of yang after the winter solstice, will make the measured pattern clear:

Head 1, Center (the phase Water): Yang ch'i, unseen,
germinates in the Yellow Palace. Good faith always resides at
the center.

Head 2, Full Circle (the phase Fire): Yang ch'i comes full circle. Divine, it returns to the beginning. Things go on to become their kinds.

Head 3, Mired (the phase Wood): Yang ch'i stirs slightly. Though stirred, it is mired [in yin]. "Mired": the difficulty attending the birth of things.

Half the cycle of eighty-one heads is devoted to the rebirth, growth, and maturation of yang, and the other half to those of yin. Each head text, after its cosmological first sentence, suggests in gnomic fashion the moral meaning for the aspirant to sagehood. The remaining parts of each head (which we need not quote) draw out the meaning of these images. In doing so, they make the book, like its exemplar, the Changes as refracted through the Ten Wings, a guide to timely, conscientious action—this time a systematic guide keyed to fine increments of time.[25]

There is a great deal more in Yang's book than that, but we are concerned with two of its accomplishments that do not draw on the Changes and its commentaries. He consistently interpreted yin and yang as types of ch'i. He integrated the conceptions of ch'i, yin-yang, and five phases throughout, making them the focus of his cosmology. For a thousand years philosophers praised the book as the greatest cosmological work of its era.

A medical book also completed the synthesis. This is not surprising when we consider how in earlier sources the theoretical conceptions of heaven and earth, the empire, and the body had grown up together.

The authors of the Inner Canon of the Yellow Emperor diagnose the ills of the human body by analyzing its imbalance with respect to the cosmic rhythms and present their methods as the revelation of a sage emperor in the golden age before history began. The book largely defines a perennial theme of classical medical discourse. Medicine, from first to last, is about ch'i: the body's endowment of it, the metabolism that extracts it from food and drink, its normal and abnormal circulation through the body, the foreign ch'i that carries disease into the system, and the ch'i of drugs and foods that rights the imbalance. The book determines the body's abnormal state by analyzing the character of its ch'i, often into two or five aspects, sometimes into specifically medical categories. The physician chooses between these to fit the task at hand. The five phases and yin-yang are now equivalent as ways to characterize ch'i.[26]

The authors of the *Inner Canon* build physiology on this basis. The book is full of such assertions as "according to the five phases, the east corresponds to 1 and 2 [in the cycle of 10] and to Wood [among the Phases] and reigns in spring. Spring corresponds to the dark blue of the sky and governs the liver functions. The liver functions belong to the Foot Attenuated Yin circulation tract."[27] We do not need to analyze such a passage to recognize in it correlations of the kind we have studied which tie normal and abnormal functions of the body to those of the cosmos.

Thus the synthesis of central concepts appeared at more or less the same time in a poetic meditation on time and change and in a collection of writings on medical doctrine. That is no coincidence. The master theme of the first is cosmic process, and that of the second, somatic process. Their authors were drawing, in stylistically very different ways, on ideas that had been converging for a long time. Other writings quickly joined and elaborated on them, such as the late chapters in *Abundant Dew on the Spring and Autumn Annals*. By A.D. 200, their way of assembling the basic notions had become standard.

Chronology of Historical Events

In Greece the archaic period ends with the start of the classical period, in the fifth century B.C. The classical period conventionally ends with the conquests of Alexander in the 330s B.C.; and the Hellenistic period, with Rome's conquest of Egypt in 30 B.C.

B.C.	CHINA	GREECE
1045–256	Chou dynasty	
722–481	Spring and Autumn period; Chou king as figurehead; shift of power to local aristocrats	
508		Cleisthenes' democratic reforms at Athens
480–221	Warring States period of the Chou dynasty; gradual consolidation of states	
479	Death of Confucius; beginning of shift among elite from military to humanistic values	

B.C.	CHINA	GREECE
431–404		Peloponnesian war between the two major Greek city-states, Athens and Sparta, and their allies, ending in the defeat of Athens
429		Death of Pericles, leader of Athenian democracy and many-times-elected general
404		Anti-democratic coup of the Thirty Tyrants at Athens
403		Restoration of democracy at Athens
367		Dionysius I, tyrant of Syracuse, dies, succeeded by Dionysius II
ca. 350	Beginning of patronage as significant institution	
341		Hermias, ruler of Atarneus and friend of Aristotle's, killed on the orders of the Persian king
338		Thebes defeated by Philip of Macedon at battle of Chaeronea
336		Philip dies and is succeeded as king of Macedon by twenty-year-old Alexander
334–323		Conquests by Alexander (including Egypt, Babylonia, Bactria, Indus Valley), ending in his death
323–		Alexander's empire is divided into kingdoms ruled by his generals and the dynasties they

B.C.	CHINA	GREECE
		founded—for example, the Ptolemies (Egypt), Seleucids (most of the rest of the old Babylonian empire), Antigonids (Macedonia), and Attalids (Pergamum)
318–307		Athens ruled by Demetrius of Phalerum (appointed by regent of Macedonia)
307		Greece "liberated" by Macedonian conqueror
256	Ch'in state abolishes Chou dynasty	
221–206	Ch'in dynasty: Ch'in conquers rival states; end of Chou dynasty and the Warring States; unification begins	
216		Hiero, ruler of Syracuse, dies
212		Romans under Marcellus take Syracuse
206–A.D.221	Han dynasty	
206–A.D.9	Western (Former) Han dynasty defeats Ch'in and rules empire	
ca. 186	Chiang-chia-shan medical, mathematical, astrological, and other manuscripts interred	
168	Ma-wang-tui medical and other manuscripts interred	
167		Romans defeat Macedonians at battle of Pydna
146		Romans sack Corinth
136	State's sponsorship restricted to five orthodox classics	

B.C.	CHINA	GREECE
124	Erudites in Grand Academy assigned pupils	
87	Onset of decentralization	
86		Romans under Sulla sack Athens
81	Imperial conference to debate state monopolies and foreign policy	
46		Julius Caesar as Pontifex Maximus reforms calendar
31		Octavian (later known as Augustus) defeats Anthony at the battle of Actium
27		Roman empire established under Augustus
ca. 7	First complete ephemeris in China	

A.D.		
9–23	Hsin dynasty of Wang Mang	
25–221	Eastern (Later) Han dynasty; it begins with enthronement of Radiantly Martial Emperor	
54–68		Nero reigns as emperor
57	Collapse of central control under way	
79	Imperial conference debates orthodox classics	
125	Thirty thousand students are enrolled in Grand Academy	
161–180		Marcus Aurelius reigns as emperor
168	Mass slaughter of high officials by court eunuchs	

A.D.	CHINA	GREECE
184	Beginning of large-scale local rebellions	
189	Mass slaughter of eunuchs by officials; governmental chaos	
190–221	Warlords wield power in China	
324		Constantine, first Roman emperor to adopt Christianity, founds Constantinople
529		Justinian forbids the teaching of pagan philosophy in Athens

Notes

INTRODUCTION

1. Montgomery 2000 provides a synoptic view of this confluence.
2. For an analysis of the resulting confusion, see Sivin 1978. The confusion has not greatly abated since.
3. Detienne 1996. The most important of our comparativist predecessors in the history of science is Joseph Needham. The dedication of this book recognizes the inspiration that Vernant and Gernet provided in their 1964 essay on comparison.
4. We use the slightly simplified version that has been widespread among historians for the past fifty years; it discards certain useless features (such as otiose umlauts) of the mid-nineteenth-century original. See Fairbank 1959, 1: 10.
5. We depart from Hucker 1985 in a very few instances where our research or that of colleagues has enabled a more exact understanding. For instance, he translates t'ai shih as "Grand Scribe" for the Chou dynasty and as "Grand Astrologer" for the Han, explaining that early in the Han this official lost his scribal functions (item 6212). We prefer to say that they became historiographic functions—thus Burton Watson's "Grand Historian" (Watson 1958). Although the latter two English versions are excellent job descriptions, Cook 1995 has confirmed that "Grand Scribe" is what the words meant. We believe that that understanding persisted in the Han despite the

change in responsibilities, and therefore so translate the title
throughout.

CHAPTER 1. *Aims and Methods*

1. For the forms of dates and other conventions, see the introduction.
2. This idea of stimulus and resonance (*kan-ying*), common in Ch'in and
   Han sources that we cite and basic to the *Huai-nan wan pi shu*,
   eventually inspired the large literature of what is generally called
   "physical studies"; see Sivin 1995d, chap. 6, pp. 190–91.

CHAPTER 2. *The Social and Institutional Framework of the Chinese Sciences*

1. *Meng-tzu*, 3A/4.
2. Lewis 1990, chap. 2.
3. Hsu Cho-yun 1965, chaps. 2–4 and pp. 175–80. We use "nobles" to
   mean aristocrats of the highest ranks.
4. Brooks and Brooks 1998, 273–74. The Brookses emphasize that
   their hypotheses about social origins based on surnames are
   tentative.
5. *Oxford English Dictionary*, senses 1, 3; Ebrey 1986, 630–33.
6. *Kuo yü*, 6: 4a; *Ku-liang chuan*, Duke Ch'eng, year 1; see also
   Huai-nan-tzu, 11 / 101 / 17. See the discussion in Brooks 1998.
7. For a typical complaint about the blurring of classes, see *Han shu*, 1B:
   65; see also Ebrey 1986, 614–16; Chü T'ung-tsu 1972, 113–22.
8. *Shih chi*, 127: 3215–22; cf. Loewe 1994a, 170–72. This chapter
   was added by Ch'u Shao-sun in the second half of the first
   century B.C.
9. Loewe 1968, 58. On the varieties and circumstances of slavery,
   see Ch'ü 1972, 135–59. On servitude and labor service down to
   the Han, see Twitchett and Loewe 1986, 89, 536–38; Loewe and
   Shaughnessy 1999, 285–86.
10. *Shih chi*, 85: 2505.
11. *Lun-yü*, 6/13, and *Meng-tzu*, 3A/5, 7B/26. See Schwartz 1985, 86;
    Graham 1989a, 31–33; Eno 1990a, 31, 190–97.
12. Ko 1991, discussing literary politics in the Han, gives examples of
    this solidarity.
13. *Li chi*, 13: 3b.
14. The importance of local popular religion in health care is a badly
    neglected topic. See Harper 1995 and the comparative remarks in

Poo 1998, 137–41. Nathan Sivin is working on a monograph on this topic.

15. Li Po-ts'ung 1990. For a famous argument that Hua T'o was not a name but an Indian epithet, see Ch'en Yin-k'o 1930.

16. Shih chi, 105: 2785ff.; Sivin 1995b.

17. Hou Han shu, chih, 25: 3572.

18. For the origins of T'ai-shih see Cook 1995. The shih in this compound is a different character from the one discussed on p. 17.

19. E.g., Han shu, 21A: 975.

20. Martzloff 1997, chap. 4.

21. Zufferey 1998 gives details, and Chang Han-tung 1984 analyzes the data.

22. Moran 1991; Biagioli 1993; Sarasohn 1993; Jardine 1998.

23. Ch'ü 1972, 127–35. K'o was not a precise label. It occasionally referred to slaves or employees but was the normal word for a client in a system of patronage. As Ebrey 1983 shows, in the second century A.D. authors applied it to relationships of dependence among the elite, based primarily on supplying access to social and political networks rather than on providing financial support. An interesting discussion of early Chinese patronage as an economic phenomenon, which oddly ignores its economics, is Eno 1997.

24. Shih chi, 75: 2354–55; Huai-nan-tzu, 12/113/5–9. Ch'ien Mu's comprehensive study of the philosophical clients of Ch'i (1956, 231–35) finds a total of only seventeen from start to finish. By "persuaders" Sinologists mean agents in pre-Han China skilled at convincing local rulers to accept a proposal, not necessarily a beneficial one.

25. Many sources, e.g., Chan-kuo ts'e, are more concerned with the advice of officials than with that of dependents.

26. Shih chi, 46: 1895.

27. Makeham 1994, 170; cf. Sivin 1995a for a detailed analysis of the arguments that led to the invention of this academy.

28. Shih chi, 75: 2353–54. Yang and Yang 1979, 76–88, translate his whole biography.

29. Knoblock 1988–94, 1: 26.

30. Shih chi, ch. 85, esp. p. 2510, trans. Yang and Yang 1979, 152–55.

31. On its coherence see Knoblock and Riegel 2000, 46–55.

282 Notes to Pages 33–41

32. *Han shu*, 44: 2145; cf. *Shih chi*, ch. 118. On kings of his father's generation who were patrons see Yü Ying-shih 1980, 85–86.
33. *Shih chi*, ch. 118; *Han shu*, ch. 4; Creel 1973, 256.
34. *Shih chi*, ch. 6, 28; *Han shu*, ch. 25A and elsewhere. High officials may have taken up some of the slack, but not obviously so. E.g., Kung-sun Hung spent much of his income entertaining friends and guests. Whether this was patronage is unclear. *Shih chi*, 112: 2951; Yü Ying-shih 1980, 84.
35. See, e.g., Nylan 1994, 145. Several instances of patronage involve official appointment.
36. See, e.g., the complex symmetries of the civil service in the classic *Chou li*.
37. Chung-kuo She-hui-k'o-hsueh Yuan 1978. For the significance of the find, see Hsia Nai 1979, 1–2.
38. Hucker 1985, 6218, 6180.
39. On the seismograph see Sleeswyk and Sivin 1983.
40. Lewis 1999, 337–38.
41. These include Tung's memorials (*Han shu*, 56: 2495–2523), excerpts from his omen interpretations in the "Treatise on Five-Phases Phenomena" in the Han history (ch. 27), and the few portions of his *Ch'un-ch'iu fan lu* that are likely to be authentic (see the bibliography).
42. Saussy 1997.
43. Berger and Luckmann 1966, 94–96.
44. Hsu Dau-lin 1970. Much popular divination was based on counting cycles of days and thus did not depend on observing celestial phenomena; see, e.g., Loewe 1994a, 218–32.
45. See the important discussion in Loewe 1994a: chap. 5.
46. Martzloff 1997, 126–35. The *Suan-shu shu* (Book of Arithmetic), a manuscript roughly three hundred years earlier and recently excavated, is less systematic; we discuss it in Chapter 5.
47. See the comprehensive treatment in Harper 1998. A typical therapeutically oriented primary source is the *Mai shu*.
48. Lewis 1999, 317–25.
49. Harper 1998, 44. Because early craftsmen are poorly documented, the collection of biographical notices in Chu Ch'i-ch'ien 1932–36 is still valuable.

50. Eno 1990a; Lewis 1999, 63–69. Hsun-tzu spent a decade or so at the end of his career in the lowly position of magistrate; see Knoblock 1988, 1: 28–31.

51. Roth 1991, 122, speculates that the book was compiled around 130 in the court of the king of Huai-nan. On the contents of the *Chuang-tzu* see Graham 1981, 27–33.

52. Pokora 1975; Forke 1907–11 (but see Pokora 1962); Ariel 1989, 1996.

53. The best statement of this important point is in Kern 2000c, 347.

54. *Lü shih ch'un-ch'iu*, 4.3/19/18–19. Cf. note 83.

55. Sivin 1995b.

56. Kern 2000b, 181–92, argues that there was no "total collapse of traditional learning" in the Ch'in period.

57. *Han shu*, 6: 159; Lewis 1999, 349–50.

58. Wang Kuo-wei 1927. On the model for local schools see Nylan 2001, 313–14.

59. Rawski 1979.

60. On the initial emphasis on ritual see Eno 1990a.

61. *Han shu*, 88: 3594.

62. *Hou Han shu*, 79A: 2546–47; Bielenstein 1980, 139–41.

63. Yü Shu-lin 1966. Tung Chung-shu, in the mid-second century, is the first said to have done so. Cheng Hsuan, who became the greatest scholiast of the second century A.D., was a disciple of Ma Jung's for three years before he laid eyes on him.

64. The most famous dispute is Hsu Shen versus Cheng Hsuan, on which see Cheng's *Po wu ching i i*. See also Ess 1993 on this topic and, on private teaching, Yü Shu-lin 1966.

65. For some examples see Brooks and Brooks 1998: apps. 1, 2. The authors are more apt to explain passages in this way than we would be, but they point to an important issue.

66. After the complete version of *Mo-tzu* disappeared not long after the second century B.C., there was no complete edition until A.D. 1445, and no commentary until 1797.

67. *Hsun-tzu*, 6/16/12, 23/88/36ff.; see Knoblock, 1988–94 1: 224; 3: 156. On other attacks by Hsun-tzu against Mencius see Moran 1983, 10, 52–58, 208–10.

68. *Hsun-tzu, ch.* 6; *Chuang-tzu, ch.* 33; *Shih chi*, 130: 3288–92. Harold Roth translates the third source in De Bary and Bloom 1999, 278–82. In

working out our views on the question of *chia*, we have garnered
much help from the work in progress of Mark Csikszentmihalyi,
Hans van Ess, Paul Rakita Goldin, Michael Lafargue, Michael Nylan,
and Kidder Smith, both in correspondence and in the Internet's
Warring States Working Group subscription list (mainly in March
1998).

69. Sivin 1995a analyzes one of these.

70. On Confucians see Wang Kuo-wei 1927, 4: 4a–20b; Twitchett and
Loewe 1986, 757. On Mohists see Graham 1985. On the *Lao-tzu*, see
Han-fei-tzu, 19 (*p'ien* 50): 11b–12a, trans. Watson 1964, 118; *Han shu*,
30: 1701–19, 1729–45. See also members of the three late Mohist
lineages abusing each other in *Chuang-tzu*, 91/33/29–31, trans.
Graham 1981, 277.

71. *Han shu*, 30: 1723; Kramers 1986, 757–59.

72. *Lun yü*, 6/26, 17/19, 5/10, 11/3. The chapter that takes up funerary
ritual is not one of the earliest ones.

73. *Huang-ti nei ching Ling shu*, 48/1/396/4–7; Sivin 1995b, 184.

74. *Han shu*, 30: 1776; *Shih chi*, 105: 2815–17; Yamada 1979, 1980; Keegan
1988, 230–31; Li Po-ts'ung 1990; Sivin 1995b. In the last source, pp.
179 et passim, the name of Ch'un-yü's teacher, Kung-sheng Yang-
ch'ing, should be Yang Ch'ing.

75. In one of the rituals "revealed" by the monarch, the officiant refers
to himself as *hsiao chao*, a term of self-reference for unordained
masters of the Celestial Masters movement (*t'ien-shih tao*; *Huang-ti chiu
ting shen tan ching chueh*, 1: 3a). The prefatory text that includes this rite
must thus be an addition from after the Han. That makes it likely
that the title of the tractate is also post-Han. For the contents of this
work see Sivin 1980.

76. Preface of Liu Hui (A.D. 263) to *Chiu chang suan shu*; Sivin 1995b.

77. E.g., Loewe 1993; Cullen 1996.

78. There is an analogous shift between the early seventeenth and mid-
eighteenth century, studied in detail in Elman 1984.

79. *Lun yü*, e.g., 19/12, 19/25. *Webster's New World Dictionary of American Usage*
discusses distinctions s.v. "argument" and "discuss." When Kroll
1985–87 argues for the importance of disputation in every aspect of
early Chinese culture, he uses the word with conspicuous vagueness.
The monographic study of the word *pien* in Lu 1991 is interesting for
its comparisons with Greece but is confined to rhetoric.

80. *Lun heng*, 39/187/25–30.

81. *Meng-tzu*, 3B/9; cf. Lau 1970, 113–15.

82. For the first story, see *Shih chi*, 76: 2370 n. 2, citing *Pieh lu*; cf. the partial translation in Graham 1978, 20–21, which gives references to other versions, not all of which involve Tsou. Kung-sun's argument is still quite controversial; specialists do not agree on whether it is about animals, classes, or words. For the second anecdote, see *Lü shih ch'un-ch'iu*, 18.5/113–14/27ff.; see the variant version in *K'ung-ts'ung-tzu*, 2: 4a–b, trans. Ariel 1989, 133–34.

83. *Lü shih ch'un-ch'iu*, 4.3/19/19–20; cf. Knoblock and Riegel 2000, 123.

84. *Hsun-tzu*, *p'ien* 6, is his critique of twelve philosophers, and *p'ien* 15 is the "debate."

85. *Yen t'ieh lun*, *Pai hu t'ung*. Nylan 1994, 142–44, tabulates conferences, and discusses them passim. See, e.g., *Yen t'ieh lun*, 9.53–54: 331–37, where the disputants repeated standard accounts of Tsou Yen's and Tung Chung-shu's positions while arguing over territorial expansion and dependence on punitive law, and the discussion in Loewe 1994b, 108–9.

86. *Han shu*, 56: 2523; cf. Queen 1996, 23.

87. *Han shu*, 27A: 1332. Aihe Wang 2000, 196–99, puts this story together and gives additional evidence of Tung's influence.

88. *Lun heng*, 45/209/13; Forke, 2: 338; *Hou Han shu*, 79A: 2553–54; Yü Shu-lin 1966, 113. The terms translated as "stumping, "explication," and "quiz" are *ho-nan*, *chieh-shuo*, and *nan-chieh*.

89. Tsien 1985, 38–45; Pan Chi-hsing 1998, 105; Loewe 1997, 165–69; Fu Ssu-nien 1930. Paper became an important medium after A.D. 265 and substantially replaced wood in the early fourth century.

90. See the survey in Shaughnessy 1997b, supplemented by Giele 1998–99.

91. Giele 1998–99; Harper 1998, 22–36. See also the discussion in Allen and Williams 2000, 118–20.

92. See Brooks and Brooks 1998, app. 1, on the accretion of texts, and Lewis 1999, 57–58, on their later compilation.

93. Keegan 1988, passim; Ma Chi-hsing 1990, 65–67; Sivin 1998.

94. Saussy 1997.

95. On this topic and on commentaries Henderson 1991 is excellent.

96. I ching, hexagram 61; cf. Shaughnessy 1997a, 158–59, for the early meaning; Wilhelm 1950, 1: 252, for late conventional interpretations; Sivin 1966 for constituents of the Changes.

97. Chou li is almost certainly later. On the Mo ching see Yates 1975 and Graham 1978. Judgments about the originality of this treatise, now part of Mo-tzu, are complicated by the fact that the Canons were composed as part of the basic teachings of an existing lineage, with canonical authority in mind. It incorporates a commentary, probably added later. In these respects, as in many others, the Mohists were anything but conventional.

98. Han shu, 30; on MSS, see n. 91.

99. See Harper 1998 for the first two and Lo 2000 for the last.

100. Ma Chi-hsing 1995, 934–40, 983–88, lists twenty-one unannotated and sixteen annotated reconstructions; Ma's own is the best available.

101. Graham 1993; 1989a, 53–64.

102. See note 41; Queen 1996, 249–54. Arbuckle 1991, 12, 34–46, makes a strong case for dating the three memorials to 130 B.C. or near it. Chin may refer to promotions rather than to appointments.

103. Shih chi, 105: 2796–2817; Hsu Han shu, chih 2, in Hou Han shu, pp. 3025–54, esp. pp. 3037–40; Sivin 1995b, 178–84; 1969, 58–62. We are grateful to Christopher Cullen for access to a draft that discusses this and the other astronomical passages.

CHAPTER 3. The Social and Institutional Framework of Greek Science

1. Diogenes Laertius 3.19.

2. Aristotle, Politics 1253b20–21, 1254a17–1255a2.

3. Finley 1968; Garlan 1988; Cartledge 1993, chap. 6.

4. Humphreys 1978, 1983; Vernant 1980; Finley 1983; Ober 1989; Osborne and Hornblower 1994.

5. Aristotle, Politics 1290b38–1291b13, 1328b33–1329a39.

6. Harrison 1968–71.

7. De Ste. Croix 1981; Finley 1985; Vidal-Naquet 1986; Cartledge, Millett, and von Reden 1998.

8. Plato, Hippias Minor 368bc.

9. Diogenes Laertius 8.1 and 2.18.

10. Havelock 1982; Harris 1989; Thomas 1992.

11. Sinclair 1988, 7.2, 169 ff.
12. Xenophon, *Memorabilia* 4.2.8 ff., tells us that Euthydemus had a substantial library of works on medicine, architecture, mathematics, and astronomy, among other subjects.
13. Plato(?), *Seventh Letter* 324b–325c.
14. Diogenes Laertius 2.6, 8.51, 63.
15. According to Plutarch, *Life of Pericles* 26.
16. Diogenes Laertius 4.46.
17. Plato, *Laws* 720a–e.
18. Diogenes Laertius 7.168 and 8.86.
19. Diogenes Laertius 5.36.
20. Plato, *Hippias Major* 282e; Isocrates, *Against the Sophists* 3.
21. Ps.-Hippocrates, *Precepts*, chap. 6, CMG, I 1, 32.6.
22. Diogenes Laertius 4.67.
23. Cohn-Haft 1956; Nutton 1988, chap. 5.
24. Saller 1982; cf. Millett 1989 for Greece.
25. Fraser 1972.
26. Diogenes Laertius 5.37 and 58.
27. Philo of Byzantium, *On Artillery Construction*, chap. 3.
28. Celsus, *On Medicine*, proem to book 1: par. 23, CML, I 21.15–21.
29. Drachmann 1948.
30. Plutarch, *Life of Marcellus* 14–17.
31. Philo of Byzantium, *On Artillery Construction*, chap. 3; Marsden 1971, 106 ff.
32. Diogenes Laertius 2.61.
33. Diogenes Laertius 7.169 and 4.38.
34. Pritchett and Neugebauer 1947; van der Waerden 1960; Meritt 1961.
35. Barton 1994.
36. Cambiano 1986; Mansfeld 1990.
37. Plato, *Parmenides* 128cd.
38. Guthrie 1962–1981, 1: 167; Burkert 1972.
39. Plato, *Republic* 600b.
40. Guthrie 1962–1981, vol. 3; Classen 1976; Lanza 1979; Kerferd 1981; Lloyd 1987, 83–102.
41. Cherniss 1945; Marrou 1956.
42. Plato(?), *Seventh Letter* 328bc.
43. Lynch 1972.
44. Too 1995.

45. Diogenes Laertius 7.2, 168, 179, and 183 f.
46. Glucker 1978; Barnes 1989.
47. Sedley 1989.
48. Diogenes Laertius 7.167 and 179. On Ariston, see Ioppolo 1980.
49. Lloyd 1983, pt. 3, chap. 2.
50. Ps.-Hippocrates, *Oath*. For one interpretation, see Edelstein 1967, 3–63.
51. Amundsen 1973, 1977.
52. Smith 1973; Lonie 1978; but contrast Thivel 1981.
53. Plato, *Protagoras* 311 bc.
54. Temkin 1973.
55. Galen, *On the Usefulness of the Parts of the Body* 7 chap. 14, 12 chap. 6, and other texts cited in Lloyd 1987, 334.
56. Edelstein 1967; Frede 1987, chaps. 13, 14; von Staden 1989; Lloyd 1996a, 35, 69 f.
57. Celsus, *On Medicine*, proem to book 1: par. 29, CML, I 22.11–13.
58. Galen, *On Sects for Beginners*, chap. 5.
59. Temkin 1973.
60. Plato, *Theaetetus* 147de, 148b.
61. Aristotle, *Nicomachean Ethics* 1096a11–17.
62. Plato, *Parmenides* 127b–d.
63. Aristotle, *Rhetoric* 1355b26–27, 1354a3–6.
64. Aristotle, *Topics* 101a26–28.
65. Aristotle, *Topics* 156b4–157a6, 161b2–3; cf. Moraux 1968.
66. Aristotle, *Topics* 161a16–b18, 164b9–12.
67. Loraux 1986.
68. Lloyd 1987, 61, 94–96.
69. Ps.-Hippocrates, *On the Art*, chap. 1, CMG, I 1 9.2 ff.
70. Ps.-Hippocrates, *On the Nature of Man*, chap. 1, CMG, I 1 3 166.2–9.
71. Plato, *Protagoras* 329b, 334e–335a.
72. Ps.-Hippocrates, *On Diseases I*: chap. 1, L VI 140.1–142.12; emphasis added.
73. Aristotle, *Rhetoric* 1419b3–4.
74. Aristotle, *On the Heavens* 294b6–10.
75. Aristotle, *Rhetoric* 1404a8–12.
76. Cambiano 1992.
77. E.g., Galen, CMG, V 10 2 2 19.5 ff., 69.19 ff.
78. Ps.-Hippocrates, *On Joints*, chap. 1, L IV 78.9–12.

79. Manuli 1983; Lloyd 1987, 105–7; Cambiano 2001.
80. Aristotle, *Metaphysics* 1093a26–28.
81. Lloyd 1991, chap. 17.
82. Cuomo 2000.
83. Sambursky 1962; Wolff 1978.
84. Pfeiffer 1968.
85. Plato, *Gorgias* 473b.

CHAPTER 4. *The Fundamental Issues of Greek Science*

1. Hesiod, *Theogony* 22–34; *Works and Days* 765–828.
2. Homer, *Odyssey* 10: 302–6.
3. Burkert 1959.
4. Plato, *Cratylus* 424d, 426d; Plato, *Theaetetus* 201de; Plato, *Timaeus* 48b, 54d–55b.
5. Aristotle, *Metaphysics* 998a25–27.
6. Proclus, *Commentary on the First Book of Euclid's Elements*, 66.7 ff.
7. Aristotle, *Metaphysics* 985a4–b3.
8. Aristotle, *Metaphysics* 983b6–11, 18–27.
9. Aristotle, *On the Soul* 411a8.
10. Kahn 1960; Guthrie 1962–81, 1: 57, 76 ff.
11. Anaxagoras, *Fragments* 1 and 6, DK.
12. Hesiod, *Theogony* 116.
13. Detienne 1996.
14. Lloyd 1991, chap. 18.
15. Ps.-Hippocrates, *On Ancient Medicine*, chap. 20, CMG, I 1, 51.6–12.
16. Plato, *Laws* 888e–889b.
17. Ps.-Hippocrates, *On the Sacred Disease*, esp. chap. 1, on which see Lloyd 1979, chap. 1.
18. Guthrie 1962–81, 3: chap. 4; Lloyd 1991, chap. 18, 425 ff.
19. Herodotus, 3.38.
20. Lloyd 1983, pt. 1, chap. 3.
21. Sambursky 1959.
22. Toomer 1984.
23. Lloyd 1991, chap. 17.
24. Epicurus, *Letter to Pythocles*, 85–88; Long and Sedley 1987, sec. 18, 90–97.
25. Long and Sedley 1987, sec. 71 and 72, 468–88.
26. Sextus Empiricus, *Outlines of Pyrrhonism* 1.98.

27. Sextus Empiricus, *Outlines of Pyrrhonism* 1.103.
28. Diogenes Laertius 7.89.
29. Aristotle, *Posterior Analytics* I, chap. 1–4, 71a1–74a3.
30. Mueller 1981.
31. Hesiod, *Theogony* 22–34.
32. Lloyd 1987, chap. 3.
33. Cf. Lloyd 1996a, chap. 5.
34. Aristotle, *Metaphysics* 986a22–26.
35. Ps.-Hippocrates, *On Regimen* 3:chap. 70, CMG, I 2 4 202.11 f.; Ps.-Hippocrates, *On Ancient Medicine*, chap. 19, CMG, I 1 50.7–9.
36. Plato, *Timaeus* 46cd.
37. Plato, *Phaedo* 98b–99b.
38. Clement, *Miscellanies* 8 9 33; and other texts in Long and Sedley 1987, sec. 55.
39. Epicurus, *Letter to Pythocles*, 96; Lucretius 5.751–70.
40. Homer, *Iliad* 19: 86–89; see Dodds 1951, chap. 1.
41. Ps.-Hippocrates, *On the Sacred Disease:* chap. 1, L VI 360.13–16.
42. Lloyd 1990, 86; cf. Lloyd 1996c, chap. 1.
43. This is documented at Lloyd 1996a, 56–58; cf. Mendell 1998.
44. Manuli 1981; Long 1989.
45. The articulation of Euclid's *Elements* and the limitations of his axiomatizations are discussed in Knorr 1981, Mueller 1981, and Netz 1999.
46. Proclus, *Commentary on the First Book of Euclid's Elements* 191.21 ff.
47. Archimedes, *On the Equilibrium of Planes* II 124.3 f.
48. Ps.-Hippocrates, *On Ancient Medicine*, chap. 9, CMG, I 1 41.20–22; cf. Lloyd 1987, 128–31, 253–57.
49. Galen, K 9.464.1–4; von Staden 1989, 354–56.
50. Compare Hankinson 1991 and Barnes 1991 with Lloyd 1996b.
51. Neugebauer 1975 sets out the history of Greek mathematical astronomy.
52. On the realist assumptions made by Greek astronomers, see Lloyd 1991, chap. 11.
53. Archimedes, *Sand-Reckoner* 1.4, II 218.7 ff.
54. Plutarch, *On the Face on the Moon*, chap. 6, 923a.
55. Ptolemy, *Planetary Hypotheses* II chaps. 6–7.
56. Ptolemy, *Syntaxis* I chap. 7: I 21.9–26.3.
57. Ptolemy, *Syntaxis* I chap. 1: I 6.11–7.4.

58. Ptolemy, Tetrabiblos I chap. 1: 4.1 ff., I chap. 2: 5.5 ff.; cf. Long 1982.
59. Lloyd 1979, 192–97.
60. Homer, Iliad 8: 13–16, 21: 195–98; Hesiod, Theogony 720–25.
61. Heraclitus, Fragment 30, DK.
62. Parmenides, Fragment 8.61, DK; cf. Fragment 1 28–32.
63. Cf. Furley 1987.
64. Hero, Pneumatics I 16.16 ff.
65. Our source for this is Simplicius, Commentary on Aristotle's Physics 467.26, quoting Eudemus's attribution to Archytas.
66. Lloyd 1966, chap. 4, analyzes these in detail.
67. Plato, Timaeus 92c; cf. 33b, 73e, 74c. See also Philebus 28c; Vlastos 1975.
68. Aristotle, Metaphysics 1076a3–4; Meteorologica I chap. 14: 351a19–353a28, esp. 351a26–31.
69. Diogenes Laertius 7.134, 138, 142–3, and other texts in Long and Sedley 1987, secs. 44, 46, and 47.
70. Diogenes Laertius 9.32; Aristotle, On the Heavens 303a14–16; and other sources on which Lloyd 1966, 248–49, comments.
71. Epicurus, Letter to Pythocles, 89.
72. Epicurus, Letter to Herodotus, 45; Long and Sedley 1987, sec. 13.
73. Heraclitus, Fragments 80 and 53, DK.
74. Anaximander, Fragment 1, DK; Empedocles, Fragment 17.27 ff., DK.
75. Lucretius 2.1093.
76. Aristotle, On the Generation of Animals 732a1–9.
77. Harrison 1968–71, 1: 108–15.
78. Hopkins 1978.
79. Lloyd 1996a, 174–79.
80. Sambursky 1959, 9 ff., 41 ff., 108 ff.
81. Sinclair 1988, chap. 5.

CHAPTER 5. The Fundamental Issues of the Chinese Sciences

1. Shang shu, 1/01/0060–0241, trans. Karlgren 1950, 3, modified; Huang-ti nei ching Ling shu, 1/1/263/1. On the odd figure of 366 for days in the year, Joseph Needham saw this as evidence that the "Yao tien" was written between the eighth and the fifth centuries B.C. but then acknowledges that even in the second millennium the tropical year

was known to be roughly 365¼ days long (Needham et al. 1954–, 3: 245 nn. b–d). Today's best estimate, that the "Yao tien" was written in the late fourth century B.C., implies that the number was a conscious archaism.

2. *Shang han lun*, preface, pp. 119, 124; *Huai-nan-tzu*, 21.226.23–24. On these two dimensions see Hadot 1990.

3. *Shih chi*, 105: 2796.

4. *Pen-ts'ao ching*. See Sivin 1987, 181, on its yin-yang and five-phases correspondences.

5. *Huai-nan-tzu*, 17.4/168/18–19; Sivin 1995c.

6. *Hsu Han shu*, in *Hou Han shu, chih* 2: 3038. See also Chia K'uei's discussion of A.D. 92 on pp. 3027–30. See Sivin 1969, 58–62, for context.

7. Graham 1989b, 512.

8. For an objection to reading "Ti" as a high god, see Eno 1990b.

9. On the cosmological aspect of military doctrine see, e.g., Sawyer 1993, 64–65, 175–76, and (on *Sun Pin ping fa*) 1995, 113–14.

10. For the very rich contents of seven such almanacs excavated in recent decades (e.g., *T'ien wen ch'i-hsiang tsa chan*), see Li Ling 2000, 43–47, 197–216.

11. See the elaborate list in Eberhard 1933 and the more useful one in Aihe Wang 2000, 110. On the audience for almanacs (*jih-shu*), see Wang, pp. 87–90.

12. Sivin 1987, 46–54.

13. On these sequences see the sources listed in note 11. On "phase," see *Oxford English Dictionary*, s.v., noun sense 2a. The production and conquest sequences appear in the almanacs of the late Warring States era; see Poo 1998, 69–101, and Wang 2000, 89. For possible precursors see *Tso chuan*, Chao 9/3 and 17/5 for 533 and 525; Legge 1872, 624, 668; and see the discussion in Li Han-san 1967, 30–31. See also Ku Chieh-kang 1930 and Sun Kuang-te 1969.

14. For a detailed discussion see Sivin 1987, 59–70.

15. The most important early systematizing treatises are *Huang-ti pa-shih-i nan ching*, *Huang-ti chia i ching*, and *Mai ching*.

16. Michael Lackner of the University of Erlangen has traced Chinese usage to the Sino-Japanese lexicon *Tetsugaku ji i* (1881; letter to Nathan Sivin, 22 Oct. 2001).

17. Graham 1989a, 383–87.

18. Chuang-tzu, ch. 29, trans. Graham 1981, 234–39. Graham, p. 28, provisionally dates this chapter ca. 200.

19. Schwartz 1985, 378–79, and Queen 1996, 228–30, argue that Tung Chung-shu both encouraged the Martial Emperor's despotic tendencies and tried to constrain them.

20. Twitchett and Loewe 1986, 190–96.

21. Zürcher 1980.

22. Lao-tzu, p'ien 1, 15, 80; cf. Lau 1982, 3, 267, 21, 287, 115–17. For a survey of the many interpretations of this book, some of which deny its mysticism, see Hardy 1998.

23. Lü shih ch'un-ch'iu, 16.6/95/24–27; cf. Knoblock and Riegel 2000, 393–94.

24. Wang (often called Wang Shu-ho), preface to Mai ching.

25. Shih chi, 26: 1256. For an analogous passage tracing divination to the "emperor" Fu-hsi, see Han shu, 27A: 1315–16. It in turn quotes a passage on this revelation in Chou i, "Hsi tz'u," sec. 11 (p. 44).

26. Tso chuan, Chao 29/app. 4, 5; 31/7; 32/3, 6 (two consultations); Ai 9/app. 3, involving three consultants; cf. Legge 1872, 819, on this passage.

27. Tso chuan, Ai 20/app. 3, for 477; cf. Legge 1872, 853. The given name here is An, but we believe this synonym of Mo is a variant reading.

28. Shih chi, 128: 3224. Loewe 1994a, 86, suggests that Wu-ti was not actually much involved in "grand political and military movements," but the evidence needs to be fully developed.

29. Kuan-tzu, 1/34/1 (ch. 3, p'ien 8), 2/65/11 (ch. 13, p'ien 36); cf. Rickett 1985–98, I: 185, II: 80; and Harold Roth in De Bary and Bloom 1999, 256–63. See also Huai-nan-tzu, 7/57/10, 10/83/10, in which the sage's "fullness is as though empty."

30. Lao-tzu, 22 (D. C. Lau 1982, 35, translates it as "empty saying").

31. Harbsmeier 1998, 247–60. The Mohists explored epistemological issues, but in this, as in other respects, philosophers and scientists ignored them.

32. Lü shih ch'un-ch'iu, 22.3/145/30–146/19; cf. Knoblock and Riegel 2000, 572–75. "What lies behind" is literally "aftertraces" (chi); see Sivin 1995c, 173–74, for a later example of its use.

33. See, e.g., Yen t'ieh lun, 28: 193.

294 Notes to Pages 213–22

34. Graham 1978.
35. There has been much controversy on this question. Two useful start-ing points for the most intense polemic, on whether the lunar lodges (hsiu) originated in India or China, are Hsia Nai 1979, chap. 3, and P'an Nai 1979.
36. Chou i, "Hsi tz'u," B/49/8, trans. Wilhelm 1: 377.
37. E.g., the famous passage in Hsun-tzu, ch. 17, trans. Knoblock 1988–94, 3: 14–22.
38. Shang shu, "Hung fan," 24: 0293–0311, 0914–0974, 0342–0358, trans. Nylan 1992, 20, 25, modified. For a complete translation see Karlgren 1950, 29–35. The Year Star is usually an invisible counter-rotating correlate of Jupiter, but if the king sees it, the term must refer to Jupiter.
39. Huai-nan-tzu, 20.210.3–5; I-yin chiu chu, 182.
40. Lü shih ch'un-ch'iu, 3.5/16/3–15; cf. Knoblock and Riegel 2000, 109–11, and Sivin in De Bary and Bloom 1999, 239–41.
41. See Sivin 1987, 133–47, on the circulation system, esp. pp. 135–36 on the three distinct terminologies for it.
42. Huang-ti nei ching Ling shu, 1/6/265/3–4. "Divine ch'i" (shen ch'i) is the normal vitality of the body. The sinews (chin, mo-chin) are the muscles, ligaments, and other fibrous tissues that operate the locomotive system of the body.
43. Sivin 1987, 95–99, 123–24, 152–61.
44. Yamada 1991. On the avoidance of dissection in autopsies in the Sung, see McKnight 1981 and Chia Ching-t'ao 1980 and, for more fragmentary evidence concerning the Ch'in, Hulsewé 1985. On Aristotle's dissections see Lloyd 1991, 180–81, 190–91.
45. Huang-ti nei ching T'ai su, 19/21/3–4, parallel passage in Huang-ti nei ching Su wen, 25/1/79/1.
46. Huang-ti nei ching T'ai Su, 3/31/3–7, parallel passage in Huang-ti nei ching Su wen, 3/1/12/1.
47. Huang-ti nei ching Su wen, 8/1–2/28/1–7; full translation in Porkert 1974: chap. 3, first item in subsections V.
48. Kuriyama 1999, 160. The doctrine of the alternating "kingship" of the visceral systems is a post-Han phenomenon. It does not appear in the Inner Canon; only the corresponding alternation of the pulses is found in the Nan ching, no. 7. For Galen, the brain was the hegemonikon, but only one (with heart and liver) of three archai.

49. Shih chi, 27: 1289. For a monograph on this chapter see Kao P'ing-tzu 1965. Sun and Kistemaker 1997 have reconstituted the sky as visible in the Han; see esp. pp. 124–35.

50. Lü shih ch'un-ch'iu, 20.5/133/10–15; cf. Knoblock and Riegel 2000, 527.

51. Huai-nan-tzu, 20/210/18–20; Lü shih ch'un-ch'iu, 1.2/2/22–25; cf. Knoblock and Riegel 2000, 66; Huai-nan-tzu, 20/214/1–13. Ho may mean either "the rivers" or "the Yellow River."

52. The classification of books in the late Western Han imperial library (Han shu, ch. 30) influenced only bibliography. On the context of this catalogue see Lewis 1999, 325–32.

53. Sivin 1969, esp. p. 11; Cullen 1993.

54. Hsu Han shu, in Hou Han shu, chih 2: 3040–43; Sivin 1969, 58–62.

55. Hsu Han shu, in Hou Han shu, chih 11: 3229; Kern 2000a and 2000c, 350, points out the importance of positive portents.

56. Sivin 1969, 6.

57. Chou pi, A6, C1–2, trans. Cullen 1996, 174, 182.

58. These insights come from Christopher Cullen, letter to Nathan Sivin, 11 Feb. 1999. In Cullen 1995 he compares Euclid and the mathematician Liu Hui (third century A.D.).

59. Martzloff 1997, 127–35. For an elaborate technical discussion, based on post-Han commentaries, of what might conceivably constitute proof in this book, see Chemla 1991, 1994, 1997a, 1997b. Complete translations of the book into English by Liu Dun, Joseph Dauben et al., and into French by Chemla and Kuo Shu-ch'un (Guo Shuchun), are under way.

60. Chiu chang suan shu, 6: 179.

61. Christopher Cullen, e-mail to Nathan Sivin, March 2001.

62. An important analysis is in Lo 2000. For the sites, see G 055, 064. The temporal sequence is cloudy because it is not possible to date the composition of most individual texts.

63. Harper 1998.

64. Sivin 1987, 102–6.

65. Pen-ts'ao ching, preface (1: 2–4); Sivin 1987, 181. For a detailed study of this drug classification, see Ma 1995, 540–600.

66. Huang ti chiu ting shen tan ching chueh, 1: 2a–b.

67. On the state rituals, see Kern 2000b; on the First Emperor's policy of suppressing books, see Petersen 1995.

APPENDIX: *Evolution of the Chinese Cosmological Synthesis*

1. Another compendium closely related to the *Tso chuan*, the *Narratives from the States* (*Kuo yü*), contains little of cosmological interest. The "Hung fan" (Great plan) chapter of the *Shang shu*, despite archaistic language, can no longer be taken as a source earlier than the *Tso chuan*. The sets of categories we discuss actually appeared earliest in the *Ku-liang chuan* (see note 3).

2. One can find precursors of Chou practices in the Shang dynasty, but its records rarely specify numbers, even of the directions (David Keightley, letter to Nathan Sivin, 28 Feb. 2000).

3. *Tso chuan*, Huan 2/6, for 710; *Ku-liang chuan*, Yin 9/3.

4. *Tso chuan*, Chao 25/2; cf. Legge 1872, 708b. Note the distinction between the six standard hues (*cai*) of robes and the five colors (*se*) as a cosmological category. On numerical categories see Needham et al. 1954–, 2: 261–65. Unlike Needham, we do not believe that they were developed by schools.

5. *Tso chuan*, Chao 1/App. 8, for 541; cf. Graham 1986, 71. What the text means by associating women with sunlight is unclear.

6. The astrologer Ts'ai Mo (p. 206), while divining, mentioned five *hsing* on earth, analogous to the three types of heavenly body, for both of which officials were appointed to be responsible. Because Ts'ai's other divinations do not mention the "five materials," his term may refer to them. *Tso chuan*, Chao 29/3/App. 4 (for 513), Chao 32/6 (for 510); Legge, 1872, 731, 741; see a similar sense in Kuo Yü, 4: 3267–70. Graham 1986 goes over in detail much the same ground as we do.

7. *Tso chuan*, Hsiang 27/5/App. 2, Chao 11/4, translated from the former; cf. Legge 1872, 534, 634. The first text to use *wu ts'ai* and identify its constituents was the *I lin* (probably first century B.C.). "Ta Yü mo," one of the forged chapters of the *Shang shu*, mentions six materials.

8. Two commentaries in *Shih chi*, 28: 1369, nn. 3 and 11, quote or epitomize lost books by Tsou. On his aim of reform see 74: 2344. In the surviving fragments there is no evidence that Tsou worked out the notion of resonant categories; see Sivin 1995a.

9. *Shih chi*, 6: 237–38; see also 28: 1366, 1368, which says that the First Emperor took this set of dynastic correlations from writings of Tsou's disciples, one or more of whom may have been among the authors

of Lü's book. For a translation and detailed discussion, see Sivin 1995a, 15–17.

10. Hsun-tzu, 6/10–14; cf. Knoblock 1988, 1: 224.

11. Meng-tzu 2A/6. Cf. Fung 1952, 1: 121–22; Watson 1963, 119–20; and Graham 1986, 76, with other citations. On the fifth hsing see P'ang P'u 1980, 1985. The pertinent texts are the first and fourth following the A version of the Ma-wang-tui Lao-tzu in Kuo-chia Wen-wu-chü 1980, and the sixth in Ching-men-shih Po-wu-kuan 1998; see also Kalinowski 1995 and 1998–99. On Hsun-tzu's own wu-hsing, see Hsun-tzu 20/48–49. There is also the similar wu ch'ang, with hsin as the fifth member, also called wu ts'ai in Liu t'ao, 3: 16a–17a, trans. Sawyer 1993, 62–63. Sawyer renders wu ts'ai as "five critical talents."

12. Other texts, possibly a little earlier or later, in which it appears cryptically but possibly in a cosmological sense are Sun-tzu ping fa and Mo-tzu.

13. 24/0157–0213, Karlgren 1948–49, 1: 233.1526; 1950, 30.5; Graham 1989a, 326; Nylan 1992.

14. Tso chuan, Ai 9/4/App. 3; Legge 1872, 819.

15. Bielenstein 1980, 143–44.

16. Lü shih ch'un-ch'iu, 1.1/1/5–8; 3.2/12/26–13/1; 13.2/64/10–15, 2.1/7/3–4; cf. Knoblock and Riegel 2000, 60, 99, 283, 79. For details on Tsou's theory see Sivin 1995a. On precursors of these monthly ordinances see Knoblock and Riegel 2000, 39–41.

17. On wu-hsing and ch'i, see Huai-nan-tzu, 20/214/3; on yin-yang and wu-hsing in military arts, see ch. 15; on topography, ch. 4.

18. Ch'un-ch'iu fan lu, ch. 13.5/61/11–12. Arbuckle 1991, 414–18, provides a valuable but unavoidably inconclusive critical discussion of this chapter. Tung's authentic astrological writings ignore the term wu-hsing. On the relation between the engendering sequence and the conquest or overcoming sequence in the five phases, see Sivin 1987, 77–79.

19. For instance, see Maruyama Masao 1962, esp. p. 14, which tabulates differences between the Kuan-tzu, two discrepant chapters of Huai-nan-tzu, and Huang-ti nei ching Su wen; see also Maruyama Toshiaki 1980. Arbuckle (see note 18) discusses the many inconsistencies in Ch'un-ch'iu fan lu.

20. Han shu, ch. 56. Aihe Wang 2000, chap. 4, provides a detailed account and analysis. See in particular her table 4.3, p. 149. On the calendar reform see Sivin 1969 and Cullen 1993.

21. The only published translation that conveys a pre-philosophical reading of the *Chou i* is Shaughnessy 1997a; see also the insights in Kunst's 1985 dissertation.

22. Two of the Wings (*Wen yen*, 2/1; *T'uan chuan*, 20/31, both printed with *Chou i*) speak in passing of yang as a kind of ch'i, and of the interaction of two ch'i, which probably (but not explicitly) refers to yin and yang. The Ten Wings do not mention *wu-hsing* or draw on their associations; two commentaries discovered at Ma-wang-tui, very similar in form to the ones that found their way into the standard set, cite them in passing as a foundation of good rule (*Erh san tzu wen*, pp. 174–77; *I chih i*, pp. 218–19). For other excavated MSS of the *Changes* and related materials, see G 043, 067.

23. *Chou i*, "Hsi-tz'u," B/45/2, trans. Shaughnessy 1997a, 205, modified; Connery 1998, 35.

24. For studies of Han interpretation of the *Changes*, see Ch'ü Wan-li 1969 and Liao Ming-ch'un et al. 1991; on the connections between the *Changes* and the medical classics, see Ho Shao-ch'u 1991; on the legacy of Han studies of the *Changes*, see Hsu Ch'in-t'ing 1975, Lu Yang 1998, and Liao et al.

25. Nylan and Sivin 1995. A full translation of the *T'ai hsuan* is in Nylan 1993. On timeliness and "situational propensities" see Jullien 1995.

26. For details see Sivin 1987, 70–80.

27. *Huang-ti nei ching Ling shu*, 41/5/379/3. On the meanings of these associations, see Sivin 1987, 59 ff., 204, 209.

Bibliography

This bibliography includes all sources cited, as well as a small number indispensable in other respects. Those wishing to look up Chinese primary sources by the English titles used in the text will find cross-references in the index.

ABBREVIATIONS

ch.: *chüan* 卷 (roll, chapter).

CMG: *Corpus Medicorum Graecorum*. Leipzig: Teubner; Berlin: Akademie-Verlag, 1908–.

CML: *Corpus Medicorum Latinorum*. Leipzig: Teubner, 1915–.

CPS: *ch'u-pan-she* 出版社 (Publishing company).

DK: *Die Fragmente der Vorsokratiker*, edited by H. Diels, revised by W. Kranz. Berlin: Weidmann, 1952.

G: site number in Giele undated.

ITCM: *I t'ung cheng mai ch'üan shu* 醫統正脈全書.

K: *Claudii Galeni opera omnia*, edited by C. G. Kühn. Leipzig: Cnobloch, 1821–33.

L: *Oeuvres complètes d'Hippocrate*, edited by E. Littré. Paris: Baillière, 1839–61.

MWT: MSS from Ma-wang-tui tomb 3 (G 064), in Kuo-chia Wen-wu-chü 1980.

TT: *Cheng-t'ung tao tsang* 政统道藏, cited by number in Schipper 1975.

GREEK AND LATIN PRIMARY SOURCES
Except where specified otherwise, we cite Greek and Latin authors by
standard editions: the Oxford Classical Texts (Oxford: Clarendon), the
Teubner (Leipzig) series, or, in default of either, the Loeb Classical Library
(Cambridge: Harvard University Press). Translations of passages cited are
our own, but we draw extensively on those in Loeb and on others cited
below. All the ancient authors we cite are included in the index.
  The following texts and translations are listed within broad categories
by chronological order of authors.

Philosophers
Presocratic philosophers and major sophists: texts in DK; translations in
  Kirk, Raven, and Schofield 1983.
Plato: text in Burnet 1900–1907.
Aristotle: text in Bekker 1831–70, by page, column, and line; translation in
  Barnes 1984.
Hellenistic philosophers: texts and translations in Long and Sedley 1987
  for preference.

Medical Writers
Greek and Latin medical writers: texts in CMG or CML when included;
  otherwise, the Hippocratic writers in L, Galen in K, specifying volume
  number.
Galen: important translations in DeLacy 1978–84; May 1968; Nutton 1979;
  C. Singer 1956; P. Singer 1997; and Walzer and Frede 1985.

Mathematical, Astronomical, Harmonic, and Mechanical Writers
Aristoxenus: text in Macran 1902; translation in Barker 1989.
Autolycus: text in Mogenet 1950; we also consult Aujac 1979.
Euclid: text in Teubner edition (Heiberg et al. 1883–1977); translation in
  Heath 1926.
Aristarchus: text and translation in Heath 1913.
Archimedes: text in Teubner edition (Heiberg 1910–72); translation in
  Heath 1912.
Philo, *On Artillery Construction (Belopoeica)*: text with translation in Marsden
  1971.
Vitruvius, *On Architecture*: text in Krohn 1912; translation in Granger
  1931–34.

Hero: text in Schmidt et al. 1899–1914; translations in Drachmann 1948, 1963; Marsden 1971.

Ptolemy, *Syntaxis mathematica*: text in Heiberg et al. 1898–1903, translation in Toomer 1984; *Planetary Hypotheses*: text in Heiberg 1907; *Tetrabiblos*: text in Hübner 1998, translation in Robbins 1940; *Optics*: text in Lejeune 1956; *Harmonics*: text in Düring 1930, translation in Barker 1989; *On the Criterion*: text and translation in Huby and Neal 1989, 179–230.

Commentators and Other Authors

Anonymus Londinensis: text in Diels 1893; translation in Jones 1947.

Proclus, *Commentary on the First Book of Euclid's Elements*: text in Friedlein 1873; translation in Morrow 1970.

Philoponus and Simplicius, the Aristotelian commentators: text in Prüssische Akademie der Wissenschaften 1882–1909.

CHINESE AND JAPANESE PRIMARY SOURCES

For classics, unless otherwise specified, we cite texts in the Harvard-Yenching or ICS Ancient Chinese Texts concordance series; for standard histories, we cite the Chung-hua Shu-chü Twenty-Four Histories. We use the earliest known title for books except when that is likely to be confusing. Texts for which we list no author are anonymous. We list only translations that we have consulted. Unless otherwise noted, we follow the arguments for datings of classical texts in Loewe 1993.

Chan-kuo ts'e 戰國策 (Intrigues of the Warring States). Compiled between 26 and 8 B.C. Shang-hai Ku-chi edition of 1978.

Chiu chang suan shu 九章算術 (later called *Chiu chang suan ching* 算經; Mathematical methods in nine chapters). Ca. A.D. 100; major revisions in late third century A.D. In Ch'ien 1963. Dating: Cullen in Loewe 1993, 19; Martzloff 1997, 129–31.

Chou i 周易 (Changes of the Chou dynasty; usually referred to as *Book of Changes*). Original text in the late ninth century B.C.; Ten Wings between the third and second centuries B.C., except "Hsu kua," in the Eastern Han. Translation: Shaughnessy 1997a.

Chou li 周禮 (Rites of the Chou dynasty). End of the first century B.C. Some scholars defend an earlier date, e.g., W. G. Boltz in Loewe 1993, 25–29.

Chou pi 周髀 (later called Chou pi suan ching 算經; Gnomon of the Chou dynasty). Between 50 B.C. and A.D. 100. In Ch'ien 1963. Text, translation, and dating: Cullen 1996.

Chuang-tzu 莊子 (Book of Chuang-tzu). Early parts by Chuang Chou 莊周? Ca. 320 to the late second century B.C. Partial translation: Graham 1981; on authorship see also Graham 1990, 283–321.

Ch'un-ch'iu 春秋 (Spring and autumn annals). After 479 B.C.

Ch'un-ch'iu fan lu 春秋繁露 (Abundant dew on the Spring and Autumn Annals). Attributed to Tung Chung-shu 董仲舒. Compiled by various hands between ca. 156 B.C. and the end of the Western Han, with only a few chapters likely to be by Tung. In Ch'un-ch'iu fan lu i cheng 春秋繁露義証. Dating: Arbuckle 1991. He points out that the present title was derived from two separate books, Ch'un-ch'iu 春秋 and Fan lu 蕃露; our translation reflects the later understanding.

Erh san tzu wen 二三子問 (The several disciples asked). Late third century B.C.? Text and translation: Shaughnessy 1997a, 168–85.

Han-fei-tzu 韓非子 (Book of Han-fei-tzu). By Han Fei. Before 233 B.C.; probably compiled (and parts written) later. In Nien-erh-tzu 廿二子.

Han shu 漢書 (Documents [i.e., history] of the [Western] Han dynasty). By Pan Ku 班固 and Pan Chao 班昭. Presented to the throne in A.D. 92; parts completed later.

Ho-kuan-tzu 鶡冠子 (Book of Ho-kuan-tzu). All or part written between 242 and 202 B.C. In Defoort 1997.

Hou Han shu 後漢書 (Documents of the Later [i.e., Eastern] Han dynasty). By Fan Yeh 范曄. Presented in A.D. 445. Treatises are from Hsu Han shu.

Hsing-te 刑德 (Punishment and virtue). Before 168 B.C. MWT. See Kalinowski 1995, 1998–99.

Hsu Han shu 續漢書 (Continuation of Documents of the [Eastern] Han dynasty). By Ssu-ma Piao 司馬彪. By A.D. 306 B.C. Lost except for treatises included in Hou Han shu. See Mansvelt Beck 1990.

Hsun-tzu 荀子 (Book of Hsun-tzu). By Hsun Ch'ing 荀卿 (or K'uang 況). Authentic chapters (no consensus on which they are) by 238 B.C., others in the Han period. Translation and dating: Knoblock 1988–94.

Huai-nan-tzu 淮南子 (Book of the King of Huai-nan). By Liu An 劉安 et al. 139 B.C. Dating: Le Blanc 1985, chap. 1.

Huai-nan wan pi shu 淮南萬畢術. (The comprehensive arts of the King of Huai-nan). Attributed to Liu An 劉安. Mid-second century B.C. In Yü han shan fang chi i shu, hsu-pien 玉函山房輯佚書續編.

Huang-ti chia i ching 黃帝甲乙經 ("A-B" canon of the Yellow Emperor). By Huang-fu Mi 皇甫謐. Between A.D. 256 and 282. In ITCM, s.v. *Huang-ti chen-chiu* 鍼灸 *chia i ching.*

Huang-ti chiu ting shen tan ching chueh 黃帝九鼎神丹經訣 (Nine-cauldron divine elixir canon of the Yellow Emperor', with explanations). Ch. 1, the canon, was written between the first century A.D. and the Six Dynasties. TT 885.

Huang-ti nei ching 黃帝內經 (Inner canon of the Yellow Emperor). Probably first century B.C. The *Ling shu* 靈樞 (Divine pivot, earlier called *Huang-ti chen ching* 黃帝鍼經 and other titles) and *Su wen* 素問 (Basic questions), edited by Wang Ping 王冰, preface A.D. 762, are cited from Jen Ying-ch'iu 1986 by *p'ien, chang,* page, and line number. *Tai su* 太素 (Grand basis), edited by Yang Shang-shan 楊上善, A.D. 656 or later, is cited from Kosoto 1981 by *p'ien,* page, and line number. Dating: Loewe 1993, 199–201; Sivin 1998.

Huang-ti pa-shi-i nan ching 黃帝八十一難經 (usually cited as *Nan ching*; Canon of eighty-one problems [in the *Inner Canon*] of the Yellow Emperor). Probably second century A.D. In *Nan ching pen i.*

I chih i 易之義 (Characteristics of the *Changes*). Late third century B.C.? Text and translation: Shaughnessy 1997a, 214–33. The translation renders the title as "The Properties of the Changes."

I ching. See *Chou i.*

I lin 易林 (Forest of the *Book of Changes*). By Chiao Kan 焦贛. Early first century B.C. TT 1475. Authenticity uncertain.

I-yin chiu chu 伊尹九主 (Yi-yin and his nine [types of] rulers). Late third century B.C.? Text and translation: Yates 1997, 180–91.

Kuan-tzu 管子 (Book of Kuan-tzu). Constituent texts written in the fourth to first centuries B.C. *Kuo-hsueh chi-pen ts'ung-shu* 國學基本叢書. Translation: Rickett 1985–98.

Ku-liang chuan 穀梁傳 (The Ku-liang tradition of interpretation of the *Spring and Autumn Annals*). Compiled between the late third and late second centuries B.C.

K'ung-ts'ung-tzu 孔叢子 (Florilegium of traditions about Confucius). By Wang Su 王肅. First half of the third century A.D. Text: *Han Wei ts'ung-shu* 漢魏叢書. Dating: Ariel 1989. Partial translation: Ariel 1989, 1996.

Kuo yü 國語 (Narratives from the states). Ca. 306 B.C.? Text: Bauer 1973. Dating: Brooks and Brooks 1998, 8.

Lao-tzu 老子 (Book of Lao-tzu). Compiled in the late third century B.C.? In Müller and Wagner 1968. Precursor texts from before ca. 320 (G 007): Ching-men-shih Po-wu-kuan 1998; see also Allan and Williams 2000.

Li chi 禮記 (Record of rites). After A.D. 79.

Liu t'ao 六韜 (Six secret teachings [lit., "bow-covers"]). Late fourth century B.C.? In *Sung-pen wu ching ch'i shu* 宋本武經七書, 1935 reprint from *Hsu ku-i ts'ung-shu* 續古逸叢書. Translation and dating: Sawyer 1993.

Lun heng 論衡 (Discourses weighed in the balance). By Wang Ch'ung 王充. Between A.D. 70 and 80? Translation: Forke 1907–11.

Lun yü 論語 (Analects). By disciples of Confucius. 479–249 B.C. Dating: Brooks and Brooks 1998.

Lü shih ch'un-ch'iu 呂氏春秋 (Springs and autumns of Master Lü). Compiled under the patronage of Lü Pu-wei 呂不韋, 239–235 B.C. Translation and dating: Knoblock and Riegel 2000.

Mai ching 脈經 (Canon of the pulsating vessels). By Wang Hsi 王熙 (or Shu-ho 叔和). Ca. A.D. 280. In ITCM.

Mai shu 脈書 (Book of the pulsating vessels). Before 186 B.C. MWT, supplemented from Chang-chia-shan MS version (G 055) in Harper 1998, app. 1.

Meng-tzu 孟子 (Book of Mencius). By Meng K'o 孟軻 et al. 320–third century B.C. Dating: Brooks and Brooks 1998, 9.

Mo ching 墨經 (The Mohist canons). Probably ca. 300 B.C. In *Mo-tzu*. Translation: Graham 1978.

Mo-tzu 墨子 (Book of Mo-tzu). Between the fourth century and ca. 100 B.C. Dating: Brooks and Brooks 1998, 9.

Nan ching pen i 難經本義 (Canon of problems: Original meanings). By Hua Shou 滑壽. A.D. 1361. Taipei: Hsuan-feng CPS, 1976.

Pai hu t'ung 白虎通 (Comprehensive discussions in the White Tiger Hall). By Pan Ku 班固. After A.D. 81. *Pai hu t'ung shu cheng* 白虎通疏证.

Pen-ts'ao ching 本草經 (Canon of materia medica). Late first or second century A.D. In Ma 1995.

Pieh lu 別錄 (Separate register). By Liu Xiang 劉向. After 26 B.C. Lost.

Po wu ching i i 駁五經異義 (Refutation of *Discrepant interpretations of the five classics*). By Cheng Hsuan 鄭玄. Late first century A.D. *Hou chih pu-tsu chai ts'ung-shu* 後知不足齋叢書.

Shang han lun 傷寒論 (Discourse on Cold Damage Disorders). See *Shang han tsa ping lun*.

Shang han tsa ping lun 傷寒雜病論 (Discourse on Cold Damage and miscellaneous disorders). By Chang Chi 張機. Between A.D. 196 and 220. For the portion that became *Shang han lun*, see Ōtsuka 1966.

Shang shu 尚書 (Book of documents). Eighth–second centuries B.C. Text: *Shang shu t'ung chien* 尚書通檢 (Peking: Shu-mu Wen-hsien CPS, 1982). Translation: Karlgren 1950.

Shen-nung pen-ts'ao 神農本草. *See Pen-ts'ao ching.*

Shih chi 史記 (Records of the Grand Scribe). By Ssu-ma T'an 司馬談 and Ssu-ma Ch'ien 司馬遷. Ca. 100 B.C.

Shu ching 書經. See *Shang shu.*

Shuo wen chieh tzu 說文解字 (Explanations of simple and compound characters). By Hsu Shen 許慎. A.D. 100. In *Shuo wen chieh tzu ku lin* 說文解字詁林.

Suan-shu shu 算數書 (Book of arithmetic). By 186 B.C. Transcription: Chiang-ling Chang-chia-shan Han chien cheng-li hsiao tsu 2000; see also P'eng Hao 2000. A second MS of the same title and nearly the same period has not yet been published (G 055, 106).

Sun Pin ping fa 孫臏兵法 (Military methods of Sun Pin). By Sun Pin 孫臏. Late fourth or early third century B.C. Peking: Wen-wu CPS, 1975. Yin-ch'ueh-shan version (G 070). Translation: Sawyer 1995.

Sun-tzu ping fa 孫子兵法 (Military methods of Sun-tzu). Between ca. 345 and ca. 272 B.C. In addition to the ICS Ancient Chinese Texts series, see G 070.

Tai hsuan 太玄 (later called *Tai hsuan ching* 太玄經; Supreme mystery). By Yang Hsiung 揚雄. Ca. 4 B.C. In *Tai hsuan chiao shih* 太玄校釋.

Tetsugaku ji i 哲學字彙 (Glossary of philosophical terms). Inoue Tetsujirō 井上哲次郎. Tokyo University, 1881.

Tien wen ch'i-hsiang tsa chan 天文氣象雜占 (Miscellaneous divinations involving celestial and meteorological portents). Before 168 B.C. MWT.

Tso chuan 左傳 (Master Tso's tradition of interpretation of the *Spring and Autumn Annals*). Attributed to Tso Ch'iu-ming 左丘明. Compiled in final form ca. 310 B.C. Dating: For the three traditions see Loewe 1993, 67–76. The historic value of the anecdotes for the years attached to them, between 722 and 464 B.C., is under debate. Brooks and Brooks (1998, 8) argue that the *Tso chuan* was written from scratch beginning ca. 350 and expanded into its final form ca. 312 and has no value as a record of earlier events. Their evidence for fourth-century language and for the date of compilation is persuasive, but they have

not disproven the use of early materials in the received version. Cf.
Pines 1997.

Tsu pi shih-i mai chiu ching 足臂十一脈灸經 (Moxibustion canon for the eleven
foot and arm vessels). Before 168 B.C. MWT. Translation: Harper 1998,
192–202.

Wu-shih-erh ping fang 五十二病方 (Formulas for fifty-two ailments). Before
168 B.C. MWT. Translation: Harper 1998, 221–304.

Yen t'ieh lun 鹽鐵論 (Discourses on salt and iron). By Huan K'uan 桓寬.
Between 79 and 49 B.C. In Wang Li-ch'i 1958.

Yin-yang shih-i mai chiu ching 陰陽十一脈灸經 (Moxibustion canon for the
eleven yin and yang vessels). Before 168 B.C. as a separate work.
Two versions in MWT and one as part of a Chang-chia-shan MS
(G 064, 055). Translation of Ma-wang-tui version A: Harper 1998,
203–12.

SECONDARY SOURCES

Allan, Sarah, and Crispin Williams, eds. 2000. *The Guodian Laozi. Proceedings
of the International Conference, Dartmouth College, May 1998.* Early China Special
Monograph Series, 5. Berkeley: Society for the Study of Early China and
the Institute of East Asian Studies, University of California.

Amundsen, Darrel W. 1973. The Liability of the Physician in Roman Law.
In *International Symposium on Society, Medicine and Law, Jerusalem, March 1972,*
edited by H. Karplus, 17–30. Amsterdam: Elsevier.

———. 1977. The Liability of the Physician in Classical Greek Legal
Theory and Practice. *Journal of the History of Medicine and Allied Sciences*
32: 172–203.

An Tso-chang 安作璋 and Hsiung T'ieh-chi 熊鐵基. 1984. *Ch'in Han kuan chih
shi kao* 秦漢官制史稿 (Draft history of the system of officials in the Ch'in
and Han periods). Chinan: Ch'i-Lu Shu She.

Arbuckle, Gary. 1991. Restoring Dong Zhongshu (BCE 195-115): An
Experiment in Historical and Philosophical Reconstruction. Ph.D. diss.,
University of British Columbia.

Ariel, Yoav. 1989. *K'ung-ts'ung-tzu. The K'ung Family Masters' Anthology. A Study and
Translation of Chapters 1–10, 12–14.* Princeton Library of Asian Translations.
Princeton University Press.

———. 1996. *K'ung Ts'ung-tzu: A Study and Translation of Chapters 15–23, with a
Reconstruction of the Hsiao Erh-ya Dictionary.* Sinica Leidensia, 35. Leiden:
E. J. Brill.

Aujac, Germaine. 1979. *Autolycos de Pitane. La sphère en mouvement, levers et couchers héliaques.* Paris: Les Belles Lettres.

Bagley, Robert, ed. 2001. *Ancient Sichuan. Treasures from a Lost Civilization.* Seattle Art Museum and Princeton University Press.

Barker, Andrew D. 1989. *Greek Musical Writings.* Vol. 2. Cambridge University Press.

————. 2000. *Scientific Method in Ptolemy's Harmonics.* Cambridge University Press.

Barnes, Jonathan. 1984. *The Complete Works of Aristotle.* Oxford: Clarendon.

————. 1989. Antiochus of Ascalon. In Griffin and Barnes 1989, 51–96.

————. 1991. Galen on Logic and Therapy. In *Galen's Method of Healing,* edited by Fridolf Kudlien and Richard J. Durling, 50–102. Leiden: E. J. Brill.

Barton, Tamsyn S. 1994. *Ancient Astrology.* London: Routledge.

Bates, Don G., ed. 1995. *Knowledge and the Scholarly Medical Traditions.* Cambridge University Press.

Bauer, Wolfgang. 1973. *A Concordance to the Kuo-yü* 國語引得. 2 vols. Research Aids Series, 11. Taipei: Chinese Materials and Research Aids Service Center.

Bekker, Immanuel. 1831–70. *Aristotelis opera.* 5 vols. Berlin: Reimer.

Berger, Peter L., and Thomas Luckmann. 1966. *The Social Construction of Reality.* Garden City, N.Y.: Doubleday.

Biagioli, Mario. 1993. *Galileo, Courtier. The Practice of Science in the Culture of Absolutism.* Science and Its Conceptual Foundations. University of Chicago Press.

Bielenstein, Hans. 1980. *The Bureaucracy of Han Times.* Cambridge Studies in Chinese History, Literature and Institutions. Cambridge University Press.

Bodde, Derk. 1936. The Attitude toward Science and Scientific Method in Ancient China. *Tien-hsia Monthly* 2: 139–60.

————. 1991. *Chinese Thought, Society, and Science. The Intellectual and Social Background of Science and Technology in China.* Honolulu: University of Hawaii Press.

Brooks, E. Bruce. 1998. Social Strata in the Middle Warring States. Warring States Working Group, Query 113 of 1 August. Photocopy, 4 pp.

Brooks, E. Bruce, and A. Taeko Brooks. 1998. *The Original Analects. Sayings of Confucius and His Successors.* Translations from the Asian Classics. New York: Columbia University Press.

Burkert, Walter. 1959. ΣΤΟΙΧΕΙΟΝ. Eine semasiologische Studie. *Philologus*
    103: 167–97.
————. 1972. *Lore and Science in Ancient Pythagoreanism*. Cambridge: Harvard
    University Press. Revised translation by E. L. Minar of *Weisheit und
    Wissenschaft: Studien zu Pythagoras, Philolaos und Plato*. Nuremberg: H. Carl,
    1962.
Burnet, John. 1900–1907. *Platonis opera*. 5 vols. Oxford: Clarendon.
Burnyeat, Myles F., ed. 1983. *The Skeptical Tradition*. Berkeley: University of
    California Press.
Cambiano, Giuseppe. 1992. La nascita dei trattati e dei manuali. In *Lo Spazio
    letterario della Grecia antica*, edited by Cambiano, L. Canfora, and D. Lanza,
    1: 525–53. Rome: Salerno.
————. 2001. Istituzioni e forme dell'attività scientifica in età ellenistica
    e romana. In *Storia della scienza*, edited by G. E. R. Lloyd et al., 1: sec. 4,
    601–17. Rome: Enciclopedia Italiana.
————, ed. 1986. *Storiografia e dossografia nella filosofia antica*. Turin: Tirrenia.
Cartledge, Paul A. 1993. *The Greeks. A Portrait of Self and Other*. Cambridge
    University Press.
Cartledge, Paul A., Paul C. Millett, and Sitta von Reden, eds. 1998. *Kosmos:
    Essays in Order, Conflict and Community in Classical Athens*. Cambridge University
    Press.
Cartledge, Paul A., Paul C. Millett, and Stephen C. Todd, eds. 1990. *Nomos:
    Essays on Athenian Law, Politics and Society*. Cambridge University Press.
Chang Han-tung 张汉东. 1984. Lun Ch'in Han po-shih chih-tu 论秦汉博士
    制度 (On the institution of the Erudite in the Ch'in and Han periods).
    In An and Hsiung 1984, 409–91.
Chemla, Karine. 1991. Theoretical Aspects of the Chinese Algorithmic
    Tradition (First to Third Century). *Historia Scientiarum* 42: 75–98.
————. 1994. Different Concepts of Equation in *The Nine Chapters on
    Mathematical Procedures* 九章算術 and in the Commentary on It by Liu Hui
    (Third Century). *Historia Scientiarum* 4. 2: 113–37.
————. 1997a. What Is at Stake in Mathematical Proofs from Third-
    Century China? *Science in Context* 10. 2: 227–51.
————. 1997b. Qu'est-ce qu'un problème dans la tradition mathématique
    de la Chine ancienne? Quelques indices glanés dans les commentaires
    rédigés entre le 3ième et le 7ième siècles au classique Han *Les neuf
    chapitres sur les procédures mathématiques*. *Extrême-Orient, Extrême-Occident* 19:
    91–126.

Ch'en Ch'i-yu 陳奇猷. 1984. Lü shih ch'un-ch'iu chiao shih 呂氏春秋校釋 (Critical edition of the Springs and Autumns of Master Lü). 4 vols. Shanghai: Hsueh-lin CPS.

Ch'en Yin-k'o 陳寅恪. 1930. San-kuo chih Ts'ao Ch'ung Hua T'o chuan yü fo-chiao ku-shih 三國志曹沖華佗傳與佛教故事 (The biographies of Ts'ao Ch'ung and Hua T'o in Records of the Three Kingdoms and Buddhist anecdotes). Ch'ing-hua hsueh-pao 清華學報 6. 1: 17–20.

Cherniss, Harold. 1945. The Riddle of the Early Academy. Berkeley: University of California Press.

Chia Ching-t'ao 賈靜濤. 1980. Chung-kuo ku-tai ti chien-yen chih-tu 中国古代的检验制度 (The ancient Chinese system of inquests). Fa-hsueh yen-chiu 法学研究 6: 59–64.

Chiang-ling Chang-chia-shan Han chien cheng-li hsiao tsu 江陵张家山汉简整理小组. 2000. Chiang-ling Chang-chia-shan Han chien Suan-shu shu shih wen 江陵张家山汉简《算数书》释文 (Transcription of the Book of Arithmetic on bamboo slips from Chang-chia-shan, Chiang-ling). Wen-wu 文物 9: 78–84. See also P'eng Hao 2000.

Ch'ien Mu 錢穆. 1956. Hsien-ch'in chu tzu hsi nien 先秦諸子繫年 (Chronological Studies of the pre-Han philosophers). 2d ed. (1st ed. 1935). 2 vols. Hong Kong University Press.

Ch'ien Pao-ts'ung 錢寶琮. 1963. Suan ching shih shu 算經十書 (Ten mathematical classics). Peking: Chung-hua Shu-chü.

Ching-men-shih Po-wu-kuan 荆门市博物馆, ed. 1998. Kuo-tien Ch'u mu chu chien 郭店楚墓竹简 (Bamboo slips from the Ch'u tomb at Kuo-tien, Hupei). Peking: Wen-wu CPS.

Chu Ch'i-ch'ien 朱啓鈐 et al. 1932–36. Che chiang lu 哲匠錄 (Records of sage craftsmen). Chung-kuo Ying-tsao Hsueh-she hui-k'an 中國營造學社彙刊, 1932, 3. 1: 123–61; 3. 2: 125–59; 3. 3: 91–120; 1933, 4. 1: 82–113; 4. 2: 60–85; 4. 3–4: 219–58; 1934, 5. 2: 74–105; 1935, 6. 2: 114–57; 1936, 6. 3: 148–82.

Ch'ü T'ung-tsu. 1972. Han Social Structure. Han Dynasty China, 1. Seattle: University of Washington Press.

Ch'ü Wan-li 屈萬里. 1969. Hsien-Ch'in Han Wei i li shu p'ing 先秦漢魏易例述評 (Critical survey of the Book of Changes from before the Han to the Wei period). Taipei: Hsueh-sheng shu-chü.

Chung-kuo She-hui-k'o-hsueh Yuan, K'ao-ku Yen-chiu-so 中国社会科学院考古研究所. 1978. Han Wei Lo-yang ch'eng nan chiao ti Ling-t'ai i-chih 汉魏洛阳城南郊的灵台遗址 (The remains of the imperial observatory of the Han and Wei dynasties in the southern outskirts of Lo-yang). *K'ao-ku* 考古 1: 54–57, pls. 1–3.

Classen, C. J., ed. 1976. *Sophistik. Wege der Forschung,* 187. Darmstadt: Wissenschaftliche Buchgesellschaft.

Cohn-Haft, Louis. 1956. *The Public Physicians of Ancient Greece.* Smith College Studies in History, 42. Northampton, Mass.: Department of History, Smith College.

Connery, Christopher Leigh. 1998. *The Empire of the Text. Writing and Authority in Early Imperial China.* Lanham, Md.: Rowman and Littlefield.

Cook, Constance. 1995. Scribes, Cooks, and Artisans. Breaking Zhou Tradition. *Early China* 20: 241–77.

Creel, Herrlee G. 1973. *Shen Pu-hai. A Chinese Political Philosopher of the Fourth Century B.C.* University of Chicago Press.

Cullen, Christopher. 1993. Motivations for Scientific Change in Ancient China: Emperor Wu and the Grand Inception Astronomical Reforms of 104 B.C. *Journal for the History of Astronomy* 24: 185–203.

———. 1995. How Can We Do the Comparative History of Mathematics? Proof in Liu Hui and the Zhou bi. *Philosophy and the History of Science. A Taiwanese Journal* 4. 1: 59–94.

———. 1996. *Astronomy and Mathematics in Ancient China: The Zhou bi suan jing.* Needham Research Institute Studies, 1. Cambridge University Press.

Cuomo, Serafina. 2000. *Pappus of Alexandria and the Mathematics of Late Antiquity.* Cambridge University Press.

De Bary, Wm. Theodore, and Irene Bloom. 1999. *Sources of Chinese Tradition.* 2d ed. Vol. 1. New York: Columbia University Press.

Defoort, Carine. 1997. *The Pheasant Cap Master (He guan zi). A Rhetorical Reading.* SUNY Series in Chinese Philosophy and Culture. Albany: State University of New York Press.

DeLacy, Phillip. 1978–84. *Galen, On the Doctrines of Hippocrates and Plato.* 3 vols. CMG V 4 1 2. Berlin: Akademie-Verlag.

De Ste. Croix, Geoffrey E. M. 1981. *The Class Struggle in the Ancient Greek World.* London: Duckworth.

Detienne, Marcel. 1996. *The Masters of the Truth in Archaic Greece.* Translated by Janet Lloyd. New York: Zone. Originally published as *Les maîtres de vérité dans la Grèce archaïque* (Paris: Maspero, 1967).

Diels, Hermann. 1893. *Anonymi Londinensis ex Aristotelis Iatricis Menoniis et aliis medicis Eclogae.* Supplementum Aristotelicum, 3.1. Berlin: Reimer.

Dillon, John M., and Anthony A. Long, eds. 1988. *The Question of "Eclecticism."* *Studies in Later Greek Philosophy.* Berkeley: University of California Press.

Dodds, E. R. 1951. *The Greeks and the Irrational.* Berkeley: University of California Press.

Drachmann, A. G. 1948. *Ktesibios, Philon and Heron. A Study in Ancient Pneumatics.* Copenhagen: Munksgaard.

————. 1963. *The Mechanical Technology of Greek and Roman Antiquity.* Copenhagen: Munksgaard.

Düring, Ingemar. 1930. *Die Harmonielehre des Klaudios Ptolemaios.* Göteborg: Elanders.

Eberhard, Wolfram. 1933. Beiträge zur kosmologischen Spekulation Chinas in der Han Zeit. *Baesslers Archiv* 16: 1–100.

Ebrey, Patricia. 1986. The Economic and Social History of Later Han. In Twitchett and Loewe 1986, 608–48.

Edelstein, Ludwig. 1967. *Ancient Medicine,* edited by Owsei Temkin and C. L. Temkin. Baltimore: Johns Hopkins University Press.

Elman, Benjamin A. 1984. *From Philosophy to Philology. Intellectual and Social Aspects of Change in Late Imperial China.* Harvard East Asian Monographs, 110. Cambridge: Council on East Asian Studies, Harvard University.

Eno, Robert. 1990a. *The Confucian Creation of Heaven: Philosophy and the Defense of Ritual Mastery.* SUNY Series in Chinese Philosophy and Culture. Albany: State University of New York Press.

————. 1990b. Was There a High God Ti in Shang Religion? *Early China* 15: 1–26.

————. 1997. Selling Sagehood: The Philosophical Marketplace in Early China. In Lieberthal, Lin, and Young 1997, 57–82.

Ess, Hans van. 1993. The Meaning of Huang-Lao in *Shiji* and *Han shu. Etudes chinoises* 12. 2: 161–77.

Fairbank, John K. 1959. *Ch'ing Documents. An Introductory Syllabus.* 2d ed. 2 vols. Cambridge: Center for East Asian Studies, Harvard University.

————, ed. 1957. *Chinese Thought and Institutions.* Comparative Studies of Cultures and Civilizations, 8. University of Chicago Press.

Finley, Moses I. 1983. *Politics in the Ancient World.* Cambridge University Press.

————. 1985. *The Ancient Economy.* 2d ed. (1st ed. 1973). London: Hogarth.

————, ed. 1968. *Slavery in Classical Antiquity.* 2d ed. (1st ed. 1960). Cambridge: Heffers.

Forke, Alfred. 1907–11. *Lun-Hêng.* Vol. 1, *Philosophical Essays of Wang Ch'ung;* Vol. 2, *Miscellaneous Essays of Wang Ch'ung.* Mitteilungen des Seminars für orientalische Sprachen, supplementary volumes, 10, 14. Shanghai: Kelly and Walsh; reprint, New York: Paragon Book Gallery, 1962.

Fraser, Peter M. 1972. *Ptolemaic Alexandria*. 3 vols. Oxford University Press.

Frede, Michael. 1987. *Essays in Ancient Philosophy*. Minneapolis: University of Minnesota Press.

Friedlein, Gottfried. 1873. *Procli Diadochi in primum Euclidis Elementorum librum commentarii*. Leipzig: Teubner.

Fu Ssu-nien 傅斯年. 1930. Chan-kuo wen-chi chung chih p'ien shih shu-t'i 戰國文集中之篇式書體 (On the form of books in collections of the Warring States period). *Kuo-li Chung-yang Yen-chiu-yuan Li-shih Yü-yen Yen-chiu-so chi-k'an* 國立中央研究院歷史語言研究所集刊 1. 2: 227–30.

Fung Yu-lan. 1952. *A History of Chinese Philosophy*. Translated by Derk Bodde. Vol. 1, *The Period of the Philosophers*. 2d ed (1st ed. 1937). Princeton University Press.

Furley, David J. 1987. *The Greek Cosmologists*. Vol. 1. Cambridge University Press.

Garlan, Yvon. 1988. *Slavery in Ancient Greece*. 2d ed. Translated by Janet Lloyd. Ithaca: Cornell University Press. Originally published as *Les Esclaves en Grèce ancienne* (Paris: Maspero, 1982).

Giele, Enno. 1998–99. Early Chinese Manuscripts, Including Addenda and Corrigenda to *New Sources of Early Chinese History: An Introduction to the Reading of Inscriptions and Manuscripts. Early China* 23–24: 247–337. Published 2001.

———. Undated. Database of Early Chinese Manuscripts. Http://humanities.uchicago.edu/easian/earlychina/resources_tools/databases/early_chinese-manuscripts/index.htm. For an introduction to the database, see Giele 1998–99. Initiated 2000.

Glucker, John. 1978. *Antiochus and the Late Academy*. Hypomnemata, 56. Göttingen: Vandenhoeck und Rupprecht.

———. 1998. Theophrastus, the Academy and the Athenian Philosophical Atmosphere. In *Theophrastus: Reappraising the Sources*, edited by Johannes M. van Ophuijsen and Marlein van Raalte, 299–316. Rutgers Studies in Classical Humanities, 8. New Brunswick, N.J.: Transaction.

Graham, A. C. 1978. *Later Mohist Logic, Ethics, and Science*. Hong Kong: Chinese University Press.

———. 1981. *Chuang-tzu. The Seven Inner Chapters and Other Writings from the Book Chuang-tzu*. London: George Allen and Unwin.

————. 1985. *Divisions in Early Mohism Reflected in the Core Chapters of Mo-tzu*. Occasional Paper and Monograph Series, 1. Singapore: Institute of East Asian Philosophies.

————. 1986. *Yin-Yang and the Nature of Correlative Thinking*. Occasional Paper and Monograph Series, 6. Singapore: Institute of East Asian Philosophies.

————. 1989a. *Disputers of the Tao. Philosophical Argument in Ancient China*. La Salle, Ill.: Open Court.

————. 1989b. A Neglected Pre-Han Philosophical Text: Ho-kuan-tzu. *Bulletin of the School of Oriental and African Studies* 52. 3: 497–532.

————. 1990. *Studies in Chinese Philosophy and Philosophical Literature*. SUNY Series in Chinese Philosophy and Culture. Albany: State University of New York Press.

————. 1993. Mo tzu. In Loewe 1993, 336–41.

Granger, Frank. 1931–34. *Vitruvius On Architecture*. Loeb Classical Library. 2 vols. London: Heinemann.

Griffin, Miriam, and Jonathan Barnes, ed. 1989. *Philosophia Togata: Essays on Philosophy and Roman Society*. Oxford University Press.

Guthrie, W. K. C. 1962–81. *A History of Greek Philosophy*. 6 vols. Cambridge University Press.

Hadot, Pierre. 1990. Forms of Life and Forms of Discourse in Ancient Philosophy. *Critical Inquiry* 16. 3: 483–505.

Hankinson, R. J. 1991. Galen on the Foundations of Science. In *Galeno: Obra Pensamiento e Influencia*, edited by J. A. López Férez, 15–29. Madrid: Universidad Nacional de Educacion a Distancia.

Hansen, Mogens H. 1987. *The Athenian Ecclesia*. 2d ed. (1st ed. 1983). Copenhagen: Museum Tusculanum Press.

————. 1991. *The Athenian Democracy in the Age of Demosthenes*. Oxford: Blackwell.

Harbsmeier, Christoph. 1998. *Language and Logic*. In Needham 1954–, 7: pt. 1. Cambridge University Press.

Hardy, Julia M. 1998. Influential Western Interpretations of the *Tao-te-ching*. In Kohn and LaFargue 1998, 165–88.

Harper, Donald J. 1995. Warring States, Ch'in, and Han Periods. In Overmyer et al. 1995, 152–60.

————. 1998. *Early Chinese Medical Literature: The Mawangdui Medical Manuscripts*. Sir Henry Wellcome Asian Series. London: Kegan Paul International.

Harris, William V. 1989. *Ancient Literacy*. Cambridge: Harvard University Press.

Harrison, Alick R. W. 1968–71. *The Laws of Athens*. 2 vols. Oxford: Clarendon.

Havelock, Eric A. 1982. *The Literate Revolution in Greece and Its Cultural Consequences*. Princeton University Press.

Heath, Thomas E. 1912. *The Works of Archimedes with the Method of Archimedes*. Cambridge University Press.

————. 1913. *Aristarchus of Samos*. Oxford: Clarendon.

————. 1926. *The Thirteen Books of Euclid's Elements*. 2d ed. (1st ed. 1908). 3 vols. Cambridge University Press.

Heiberg, J. L. 1907. [Ptolemy,] *Opera astronomica minora*. Leipzig: Teubner. Includes *Tetrabiblos*.

————. 1910–72. *Archimedis opera omnia*, revised by E. S. Stamatis. 4 vols. Leipzig: Teubner.

Heiberg, J. L., et al. 1883–1977. [Euclid,] *Elementa*, revised by E. S. Stamatis. 5 vols. in 6. Leipzig: Teubner.

————. 1898–1903. [Ptolemy,] *Syntaxis mathematica*. 2 vols. Leipzig: Teubner.

Henderson, John B. 1991. *Scripture, Canon and Commentary. A Comparison of Confucian and Western Exegesis*. Princeton University Press.

Ho Shao-ch'u 何少初. 1991. *Ku-tai ming i chieh Chou i* 古代名医解周易 (Eminent doctors of ancient times explicate the Book of Changes). Peking: Chung-kuo I-yao K'o-chi CPS.

Hopkins, M. Keith. 1978. *Conquerors and Slaves*. Sociological Studies in Roman History, 1. Cambridge University Press.

Hsia Nai 夏鼐. 1977. *K'ao-ku-hsueh ho k'o-chi-shih—Tsui chin wo-kuo yu kuan k'o-chi-shih ti k'ao-ku hsin fa-hsien* 考古学和科技史—最近我国有关科技史的考古新发现 (Archeology and the history of science and technology. Notes on recent archeological finds in China concerning the history of science and technology). *K'ao-ku* 考古 2: 81–91. Reprinted in Hsia 1979, 1–14.

————. 1979. *K'ao-ku-hsueh ho k'o-chi-shih* 考古学和科技史 (Archaeology and the history of science and technology). Peking: K'o-hsueh CPS.

Hsu Ch'in-t'ing 徐芹庭. 1975. *Liang Han shih-liu-chia I chu ch'an wei* 两漢十六家易注闡微 (Elucidation of subtleties in the commentaries of sixteen Han experts on the Book of Changes). Taipei: Shih-chieh T'u-shu Kung-ssu.

Hsu, Cho-yun. 1965. *Ancient China in Transition. An Analysis of Social Mobility, 722-222 B.C.* Stanford University Press.

Hsu Dau-lin. 1970. Crime and Cosmic Order. *Harvard Journal of Asiatic Studies* 30: 111–25.

Hübner, Wolfgang, ed. 1998. *Apotelesmatika.* In *Claudii Ptolemaei Opera quae exstant omnia,* by F. Boll and A. Boll, 3: pt. 1. Revision of 1898 Teubner edition. Stuttgart: Teubner.

Huby, Pamela M., and Gordon C. Neal, eds. 1989. *The Criterion of Truth.* Liverpool University Press.

Hucker, Charles O. 1985. *A Dictionary of Official Titles in Imperial China.* Stanford University Press.

Huffman, Carl A. 1993. *Philolaus of Croton.* Cambridge University Press.

Hulsewé, A. F. P. 1985. *Remnants of Ch'in Law. An Annotated Translation of the Ch'in Legal and Administrative Rules of the 3rd Century B.C., Discovered in Yun-meng Prefecture, Hu-pei Province, 1975.* Sinica Leidensia, 17. Leiden: E. J. Brill.

Humphreys, Sally C. 1978. *Anthropology and the Greeks.* London: Routledge and Kegan Paul.

———. 1983. *The Family, Women and Death.* London: Routledge and Kegan Paul.

Ioppolo, Anna-Maria. 1980. *Aristone di Chio e lo stoicismo antico.* Naples: Bibliopolis.

Jardine, Nicholas. 1998. The Places of Astronomy in Early-modern Culture. *Journal for the History of Astronomy* 29. 1: 49–62.

Jen Ying-ch'iu 任應秋 et al., eds. 1986. *Huang-ti nei ching chang-chü so-yin* 黄帝内經章句索引 (Phrase index to the *Inner Canon of the Yellow Emperor*). Peking: Jen-min Wei-sheng CPS.

Jones, A. H. M. 1964. *The Later Roman Empire 284–602. A Social, Economic, and Administrative Survey.* 2 vols. Oxford: Blackwell.

Jones, W. H. S. 1947. *The Medical Writings of Anonymus Londinensis.* Cambridge University Press.

Jullien, François. 1995. *The Propensity of Things.* Translated by Janet Lloyd. New York: Zone. Originally published as *La propension des choses: Pour une histoire de l'efficacité en Chine* (Paris: Editions du Seuil, 1992).

Kahn, Charles H. 1960. *Anaximander and the Origins of Greek Cosmology.* New York: Columbia University Press.

Kalinowski, Marc. 1995. *Ma-wang-tui po shu Hsing te shi tan* 马王堆帛书《刑德》试探 (An exploratory inquiry into the Ma-wang-tui silk MS "Hsing-te"). *Hua hsueh* 华学 1: 82–110.

————. 1998–99. The Xingde Texts from Mawangdui. Early China 23–24: 125–202.

Kao P'ing-tzu 高平子. 1965. Shih chi T'ien-kuan shu chin chu 史記天官書今註 ("Treatise on Celestial Offices" of the Records of the Grand Scribe, translated into modern Chinese and annotated). Taipei: Chung-hua Ts'ung-shu Pien-shen Wei-yuan-hui.

Karlgren, Bernhard. 1948–49. Glosses on the Book of Documents. Bulletin of the Museum of Far Eastern Antiquities (Stockholm) 20–21: entire. Reprint in 1 vol. Stockholm: The Museum, 1970.

————. 1950. The Book of Documents. Bulletin of the Museum of Far Eastern Antiquities (Stockholm) 22: 1–81. Translation of Shang shu.

Keegan, David. 1988. Huang-ti nei-ching. The Structure of the Compilation, the Significance of the Structure. Ph.D. diss., University of California, Berkeley.

Kerferd, George B. 1981. The Sophistic Movement. Cambridge University Press.

Kern, Martin. 2000a. Religious Anxiety and Political Interest in Western Han Omen Interpretation: The Case of the Han Wudi 漢武帝 Period (141–87 B.C.). Chūgoku shigaku 中國史學 10: 1–31.

————. 2000b. The Stele Inscriptions of Ch'in Shih-huang: Text and Ritual in Early Chinese Imperial Representation. American Oriental Series, 85. New Haven: American Oriental Society.

————. 2000c. Review of Writing and Authority in Early China, by Mark Edward Lewis. China Review International 7. 2: 336–76.

Kirk, Geoffrey S., John Raven, and Malcolm Schofield. 1983. The Presocratic Philosophers. 2d ed. (1st ed. 1957). Cambridge University Press.

Knoblock, John. 1988–94. Xunzi. A Translation and Study of the Complete Works. 3 vols. Stanford University Press. Translation of Hsun-tzu.

Knoblock, John, and Jeffrey Riegel. 2000. The Annals of Lü Buwei. Stanford University Press. Translation and study of Lü shih ch'un-ch'iu.

Knorr, Wilbur R. 1975. The Evolution of the Euclidean Elements. Dordrecht: Reidel.

————. 1981. On the Early History of Axiomatics: The Interaction of Mathematics and Philosophy in Greek Antiquity. In Theory Change, Ancient Axiomatics and Galileo's Methodology, edited by J. Hintikka, D. Gruender, and E. Agazzi, 1: 145–86. Dordrecht: Reidel.

Ko, Susan Schor. 1991. Literary Politics in the Han. Ph.D. diss., Yale University.

Kohn, Livia, and Michael LaFargue, eds. 1998. *Lao-tzu and the Tao-te-ching*. Albany: State University of New York Press. Essays on the book.

Kosoto Hiroshi 小曽戸洋 et al., eds. 1981. *Tōyō igaku zempon sōsho* 東洋醫學善本叢書 (Collected rare books on Oriental medicine). Osaka: Tōyō Igaku Kenkyūkai.

Kramers, Robert P. 1986. The Development of the Confucian Schools. In Twitchett and Loewe 1986, 747–65.

Krohn, Friedrich. 1912. *Vitruvii De architectura, libri decem*. Leipzig: Teubner.

Kroll, J. L. 1985–87. Disputation in Ancient Chinese Culture. *Early China* 11–12: 118–45.

Ku Chieh-kang 顧頡剛. 1930. Wu-te chung-shih shuo hsia ti cheng-chih ho li-shih 五德終始説下的政治和歷史 (Politics and history under the doctrine of the five-powers cycle). *Ch'ing-hua hsueh-pao* 清華學報 6: 71–268.

Kunst, Richard Alan. 1985. The Original "Yijing": A Text, Phonetic Transcription, Translation, and Indexes, with Sample Glosses. Ph.D. diss., University of California, Berkeley. Study and partial translation of Chou i.

Kuo-chia Wen-wu-chü, Ku-wen-hsien Yen-chiu-shih 國家文物局古文獻研究室, ed. 1980. *Ma-wang-tui Han mu po shu* 馬王堆漢墓帛書 (Silk manuscripts from the Han tomb at Ma-wang-tui). Peking: Wen-wu CPS.

Kuriyama, Shigehisa. 1999. *The Expressiveness of the Body and the Divergence of Greek and Chinese Medicine*. New York: Zone.

Lanza, Diego. 1979. *Lingua e discorso nell'Atene delle professioni*. Naples: Liguori.

Lau, D. C., trans. 1970. *Mencius*. Harmondsworth, England: Penguin. Translation of *Meng-tzu*.

———. 1982. *Chinese Classics. Tao Te Ching*. 2d ed. Hong Kong: Chinese University Press. Translation of Lao-tzu.

Le Blanc, Charles. 1985. *Huai-nan Tzu. Philosophical Synthesis in Early Han Thought. The Idea of Resonance (Kan-Ying). With a Translation and Analysis of Chapter Six*. Hong Kong University Press.

Legge, James. 1872. *The Chinese Classics. With a Translation, Critical and Exegetical Notes, Prolegomena, and Copious Indexes*. Vol. 5, *The Ch'un Ts'ew, with the Tso Chuen*. 2 fascicles. Hong Kong: Lane Crawford. Translation of Ch'un-ch'iu and *Tso chuan*.

Lejeune, Albert. 1956. *L'Optique de Claude Ptolémée*. Louvain: Publications Universitaires.

Lewis, Mark Edward. 1990. *Sanctioned Violence in Early China.* SUNY Series in
   Chinese Philosophy and Culture. Albany: State University of New York
   Press.
———. 1999. *Writing and Authority in Early China.* SUNY Series in Chinese
   Philosophy and Culture. Albany: State University of New York Press.
Li Chien-min 李建民. 2001. Chou Ch'in mai hsueh ti wang-kuan
   yuan-liu 周秦脈學的王官源流 (History of [the trope] of royal officials
   in pulse studies of the Chou and Ch'in periods). *K'o-chi i-liao yü she-hui*
   科技醫療與社會 1: 137–89.
Li Han-san 李漢三. 1967. *Hsien-Ch'in liang Han chih yin-yang wu-hsing hsueh-shuo*
   先秦兩漢之陰陽五行學說 (The yin-yang and five-phases doctrines of
   pre-Han and Han China). Taipei: Chung-ting Wen-hua Ch'u-pan
   Kung-ssu.
Li Ling 李零. 2000. *Chung-kuo fang-shu k'ao* 中国方术考 (Studies of Chinese
   technical methods). Peking: Tung-fang CPS. Mainly on divination and
   the arts of long life and immortality.
Li Po-ts'ung 李伯聪. 1990. Pien Ch'ueh ho Pien Ch'ueh hsueh-p'ai
   yen-chiu 扁鹊和扁鹊学派研究 (Studies of Pien Ch'ueh and the school of
   Pien Ch'ueh). Sian: Shan-hsi K'o-hsueh Chi-shu CPS.
Liao Ming-ch'un 廖名春 et al. 1991. *Chou i yen-chiu shih* 周易研究史 (History
   of studies of the *Book of Changes*). Changsha: Hunan CPS.
Lieberthal, Kenneth G., Shuen-fu Lin, and Ernest P. Young, ed. 1997.
   Constructing China. The Interaction of Culture and Economics. Ann Arbor: Center
   for Chinese Studies, University of Michigan.
Lloyd, Geoffrey E. R. 1966. *Polarity and Analogy.* Cambridge University
   Press.
———. 1979. *Magic, Reason and Experience.* Cambridge University Press.
———. 1983. *Science, Folklore and Ideology.* Cambridge University Press.
———. 1987. *The Revolutions of Wisdom.* Berkeley: University of California
   Press.
———. 1990. *Demystifying Mentalities.* Cambridge University Press.
———. 1991. *Methods and Problems in Greek Science.* Cambridge University
   Press.
———. 1996a. *Adversaries and Authorities.* Cambridge University Press.
———. 1996b. Theories and Practices of Demonstration in Galen. In
   Rationality in Greek Thought, edited by Michael Frede and Gisela Striker,
   255–77. Oxford: Clarendon Press.
———. 1996c. *Aristotelian Explorations.* Cambridge University Press.

Lo, Vivienne. 2000. Crossing the Neiguan 內關: A Nei/wai 內外
"Inner/Outer" Distinction in Early Chinese Medicine. *East Asian Science,
Technology, and Medicine* 17: 15–65. Published 2001.

Loewe, Michael. 1968. *Everyday Life in Early Imperial China during the Han Period,
202 BC–AD 220*. London: Batsford.

———. 1988. The Oracles of the Clouds and the Winds. *Bulletin of the School
of Oriental and African Studies* 51. 3: 500–520.

———. 1994a. *Divination, Mythology and Monarchy in Han China*. University
of Cambridge Oriental Publications, 48. Cambridge University Press.
Selected essays.

———. 1994b. *Crisis and Conflict in Han China, 104 BC to AD 9*. London:
George Allen and Unwin.

———. 1997. Wood and Bamboo Administrative Documents of the Han
Period. In Shaughnessy 1997b, 160–92.

———, ed. 1993. *Early Chinese Texts. A Bibliographical Guide*. Early China Special
Monograph Series, 2. Berkeley: Society for the Study of Early China and
the Institute of East Asian Studies, University of California.

Loewe, Michael, and Edward L. Shaughnessy, eds. 1999. *The Cambridge
History of Ancient China. From the Origins of Civilization to 221 B.C.* Cambridge
University Press.

Long, Anthony A. 1982. Astrology: Arguments Pro and Contra. In *Science
and Speculation*, edited by Jonathan Barnes et al., 165–92. Cambridge
University Press.

———. 1989. Ptolemy on the Criterion: An Epistemology for the
Practising Scientist. In Huby and Neal 1989, 157–78.

Long, Anthony A., and David N. Sedley. 1987. *The Hellenistic Philosophers*. 2
vols. Cambridge University Press.

Lonie, Iain M. 1978. Cos versus Cnidus and the Historians. *History of Science*
16: 42–75, 77–92.

Loraux, Nicole. 1986. *The Invention of Athens*. Translated by A. Sheridan.
Cambridge: Harvard University Press. Originally published as *L'Invention
d'Athènes* (Paris: Mouton, 1981).

Lu, Xing. 1991. Recovering the Past. Identification of Chinese Senses of
pien and a Comparison of *pien* to Greek Senses of Rhetoric in the Fifth
and Third Centuries BCE. Ph.D. diss., University of Oregon.

Lu Yang 卢央. 1998. *Ching Fang p'ing chuan* 京房评传 (Critical biography
of Ching Fang). Chung-kuo ssu-hsiang-chia p'ing-chuan ts'ung-shu
中国思想家评传丛书 (Critical biography series of Chinese

thinkers). Nanjing University Press. Detailed study of Ching's methods.

Lynch, John Patrick. 1972. *Aristotle's School*. Berkeley: University of California Press.

Ma Chi-hsing 马继兴. 1990. *Chung-i wen-hsien-hsueh* 中医文献学 (The study of Chinese medical literature). Shanghai: Shang-hai K'o-hsueh Chi-shu CPS.

——— 馬繼興, ed. 1995. *Shen-nung pen-ts'ao ching chi chu* 神農本草經輯注 (Reconstituted and annotated *Divine Husbandman's Materia Medica*). Peking: Jen-min Wei-sheng CPS.

Macran, Henry S. 1902. *The Harmonics of Aristoxenus*. Oxford: Clarendon Press.

Makeham, John. 1994. *Name and Actuality in Early Chinese Thought*. SUNY Series in Chinese Philosophy and Culture. Albany: State University of New York Press.

Manetti, Daniela. 1998. Commenti ed enciclopedie. In *I Greci*, edited by Salvatore Settis, 2. 3: 1199–1220. Turin: Einaudi.

Mansfeld, Jaap. 1990. *Studies in the Historiography of Greek Philosophy*. Assen: Van Gorcum.

Mansvelt Beck, B. J. 1990. *The Treatises of Later Han. Their Author, Sources, Contents and Place in Chinese Historiography*. Leiden: E. J. Brill. Study of the *Hsu Han shu*.

Manuli, Paola. 1981. Claudio Tolomeo: Il criterio e il principio. *Rivista critica di storia della filosofia* 36: 64–88.

———. 1983. Lo stile del commento. In *Formes de pensée dans la collection hippocratique*, edited by F. Lasserre and P. Mudry, 471–82. Geneva: Droz.

Marrou, Henri I. 1956. *A History of Education in Antiquity*. Translated by G. Lamb. London: Sheed and Ward. Originally published as *Une histoire de l'education dans l'antiquité* (Paris: Editions du Seuil, 1948).

Marsden, Eric W. 1971. *Greek and Roman Artillery, Technical Treatises*. Oxford: Clarendon.

Martzloff, Jean-Claude. 1997. *A History of Chinese Mathematics*. Translated by Stephen L. Wilson. Berlin: Springer-Verlag. Originally published as *Histoire des mathématiques chinoises* (Paris: Masson, 1988).

Maruyama Masao 丸山昌朗. 1962. Somon, Reisu ni okeru on'yō gogyō setsu no igi 素問靈樞に於ける陰陽五行説の意義 (The significance of the yin-yang and five-phases doctrines in the *Basic Questions* and *Divine Pivot*). *Nihon tōyō igaku kaishi* 日本東洋醫學會誌 13. 1: 13–17.

Maruyama Toshiaki 丸山敏秋. 1980. Chūgoku kodai igaku ni okeru gogyōron no kōsatsu—shinkyūkei igaku ni tsuite 中國古代醫學における五行論の考察—針灸系醫學について (An investigation of five-phases doctrine in ancient Chinese medicine: Acupuncture and moxibustion medicine). *Rinri shisō kenkyū* 倫理思想研究 5: 1–18.

May, Margaret T. 1968. *Galen on the Usefulness of the Parts of the Body*. 2 vols. Ithaca: Cornell University Press.

McKnight, Brian E. 1981. *The Washing Away of Wrongs: Forensic Medicine in Thirteenth-Century China*. Science, Medicine, and Technology in East Asia, 1. Ann Arbor: Center for Chinese Studies, University of Michigan.

Mendell, Henry. 1998. Making Sense of Aristotelian Demonstration. *Oxford Studies in Ancient Philosophy* 16: 161–225.

Meritt, Benjamin D. 1961. *The Athenian Year*. Berkeley: University of California Press.

Millett, Paul C. 1989. Patronage and Its Avoidance in Classical Athens. In *Patronage in Ancient Society*, edited by Andrew Wallace-Hadrill, 15–47. London: Routledge.

Mogenet, Joseph. 1950. *Autolycus de Pitane*. Université de Louvain, Receuil de Travaux d'Histoire et de Philologie, 37. Louvain: Bibliothèque de l'Université.

Montgomery, Scott L. 2000. *Science in Translation. Movements of Knowledge through Cultures and Time*. University of Chicago Press.

Moran, Bruce, ed. 1991. *Patronage and Institutions. Science, Technology, and Medicine at the European Court, 1500–1750*. Rochester, N.Y.: Boydell Press.

Moran, Patrick E. 1983. Key Philosophical and Cosmological Terms in Chinese Philosophy: Their History from the Beginning to Chu Hsi (1130–1200). Ph.D. diss., University of Pennsylvania.

Moraux, Paul. 1968. La joute dialectique d'après le huitième livre des *Topiques*. In *Aristotle on Dialectic*, edited by Gwilym E. L. Owen, 277–311. Oxford: Clarendon.

Morrow, Glenn R. 1970. *Proclus. A Commentary on the First Book of Euclid's Elements*. Princeton University Press.

Mueller, Ian. 1981. *Philosophy of Mathematics and Deductive Structure in Euclid's Elements*. Cambridge: Harvard University Press.

Müller, C. C., and R. G. Wagner. 1968. *Konkordanz zum Lao-tzu*. Publikationen der Fachschaft Sinologie München, 19. Seminars für Ostasiatische Sprach- und Kulturwissenschaft der Universität München. Based on *Chu tzu chi ch'eng* 諸子集成 ed.

Needham, Joseph, et al. 1954–. *Science and Civilisation in China.* 21 vols. to date. Cambridge University Press.

Netz, Reviel. 1999. *The Shaping of Deduction in Greek Mathematics.* Cambridge University Press.

Neugebauer, Otto. 1975. *A History of Ancient Mathematical Astronomy.* 3 vols. Berlin: Springer.

Nutton, Vivian. 1979. *Galen on Prognosis.* CMG V 8 1. Berlin: Akademie-Verlag.

————. 1988. *From Democedes to Harvey.* London: Variorum.

Nylan, Michael. 1992. *The Shifting Center. The Original "Great Plan" and Later Readings.* Monumenta Serica Monograph Series, 24. Nettetal: Steyler Verlag.

————. 1993. *The Canon of Supreme Mystery.* SUNY Series in Chinese Philosophy and Culture. Albany: State University of New York Press. Translation of *T'ai hsuan.*

————. 1994. The *chin wen / ku wen* Controversy in Han Times. *T'oung Pao* 80: 83–145.

————. 2001. The Legacies of the Chengdu Plain. In Bagley 2001, 309–25.

Nylan, Michael, and Nathan Sivin. 1995. The First Neo-Confucianism. An Introduction to the "Canon of Supreme Mystery" (*T'ai hsuan ching,* ca. 4 B.C.). In Sivin 1995e, chap. 3. Revised version of 1987 article.

Ober, Josiah. 1989. *Mass and Elite in Democratic Athens.* Princeton University Press.

Osborne, Robin, and Simon Hornblower, eds. 1994. *Ritual, Finance, Politics. Athenian Democratic Accounts Presented to David Lewis.* Oxford University Press.

Ōtsuka Keisetsu 大塚敬節. 1966. *Rinsō ōyō Shōkanron gaisetsu* 臨床應用傷寒論概説 (Introduction to the *Shang han lun* for clinical use). Osaka: Sōgensha. Includes critical text.

Overmyer, Daniel, et al. 1995. Chinese Religions: The State of the Field. Pt. 1, Early Religious Traditions: The Neolithic Period through the Han Dynasty (ca. 4000 B.C.E. to 220 C.E.). *Journal of Asian Studies* 54. 1: 124–60.

P'an Chi-hsing 潘吉星. 1998. *Chung-kuo k'o-hsueh chi-shu shih. Tsao-chih yü yin-shua chüan* 中国科学技术史。造纸与印刷卷 (History of Chinese science and technology. Papermaking and printing). Peking: K'o-hsueh CPS.

P'an Nai 潘鼐. 1979. Wo-kuo tsao-ch'i ti erh-shih-pa hsiu kuan-ts'e chi ch'i shih-tai k'ao 我国早期的二十八宿观测及其时代考 (On early Chinese observations of the twenty-eight lunar lodges and their periods). Chung-hua wen shih lun-ts'ung 中华文史论丛 11: 137–82.

P'ang P'u 庞朴. 1980. Po shu wu-hsing p'ien yen-chiu 帛书五行篇研究 (Studies of the silk manuscript text on the five activities). Chinan: Ch'i-Lu Shu She.

———. 1985. Origins of the Yin-Yang and Five Elements Concepts. Social Sciences in China, Spring: 91–131.

P'eng Hao 彭浩. 2000. Chung-kuo tsui tsao ti shu-hsueh chu-tso Suan shu shu 中国最早的数学著作《算数书》(On the Book of Arithmetic, China's earliest mathematics book). Wen-wu 9: 85–90.

Petersen, Jens Østergard. 1995. Which Books Did the First Emperor of Ch'in Burn? On the Meaning of pai-chia in Early Chinese Sources. Monumenta Serica 43. 1: 1–52.

Pfeiffer, Rudolf. 1968. History of Classical Scholarship from the Beginnings to the End of the Hellenistic Age. Oxford: Clarendon.

Pines, Yuri. 1997. Intellectual Change in the Chunqiu Period: The Reliability of the Speeches in the Zuo zhuan as Sources of Chunqiu Intellectual History. Early China 22: 77–132.

Pokora, Timotheus. 1962. The Necessity of a More Thorough Study of Philosopher Wang Ch'ung and His Predecessors. Archiv Orientalni 30. 2: 231–57.

———. 1975. Hsin-lun (New Treatise) and Other Writings by Huan Tan (43 B.C.–29 A.D.). Michigan Papers in Chinese Studies, 20. Ann Arbor: Center for Chinese Studies, University of Michigan.

Poo Mu-chou. 1998. In Search of Personal Welfare. A View of Ancient Chinese Religion. SUNY Series in Chinese Philosophy and Culture. Albany: State University of New York Press.

Porkert, Manfred. 1974. The Theoretical Foundations of Chinese Medicine: Systems of Correspondence. MIT East Asian Science Series, 3. Cambridge: MIT Press.

Pritchett, W. Kendrick, and Otto Neugebauer. 1947. The Calendars of Athens. Cambridge: Harvard University Press.

Prüssische Akademie der Wissenschaften. 1882–1909. Commentaria in Aristotelem graeca. 23 vols. in 48. Berlin: Reimer.

Queen, Sarah. 1996. From Chronicle to Canon: The Hermeneutics of the Spring and Autumn, according to Tung Chung-shu. Cambridge Studies in Chinese History, Literature and Institutions, 20. Cambridge University Press.

Rawski, Evelyn S. 1979. Education and Popular Culture in Ch'ing China. Ann Arbor: University of Michigan Press.

Rickett, W. Allyn. 1985–98. *Guanzi. Political, Economic, and Philosophical Essays from Early China*. 2 vols. Princeton University Press. Translation of *Kuan-tzu*.

Robbins, Frank E. 1940. *Tetrabiblos*. Loeb Classical Library. London: Heinemann.

Rosemont, Henry, Jr., ed. 1991. *Chinese Texts and Philosophical Contexts. Essays Dedicated to Angus C. Graham*. Critics and Their Critics, 1. La Salle, Ill.: Open Court.

Roth, Harold. 1991. Who Compiled the *Chuang Tzu*? In Rosemont 1991, 79–128.

Saller, Richard. 1982. *Personal Patronage under the Early Empire*. Cambridge University Press.

Sambursky, Samuel. 1959. *The Physics of the Stoics*. London: Routledge and Kegan Paul.

————. 1962. *The Physical World of Late Antiquity*. London: Routledge and Kegan Paul.

Sarasohn, Lisa T. 1993. Nicolas-Claude Fabri de Peiresc and the Patronage of the New Science in the Seventeenth Century. *Isis* 84: 70–90.

Saussy, Haun. 1997. The Unreliable Anthologists. Liu Xiang and Liu Xin in Historical Perspective. Unpublished paper for Association for Asian Studies annual meeting, Chicago.

Sawyer, Ralph D., with Mei-chün Sawyer, trans. 1993. *The Seven Military Classics of Ancient China*. History and Warfare. Boulder, Colo.: Westview Press.

————. 1995. *Sun Pin. Military Methods*. History and Warfare. Boulder, Colo.: Westview Press.

Schipper, K. M. 1975. *Concordance du Tao-tsang. Titres des ouvrages*. Publications de l'Ecole Française d'Extrême-Orient, 102. Paris: Ecole Française d'Extrême-Orient.

Schmidt, Wilhelm, et al. 1899–1914. *Heronis Alexandrini opera quae supersunt omnia*. 5 vols. Leipzig: Teubner.

Schwartz, Benjamin I. 1957. The Intellectual History of China. Preliminary Reflections. In Fairbank 1957, 15–30.

————. 1985. *The World of Thought in Ancient China*. Cambridge: Harvard University Press.

Sedley, David N. 1989. Philosophical Allegiance in the Greco-Roman World. In Griffin and Barnes 1989, 97–119.

Shaughnessy, Edward L. 1997a. *I Ching. The Classic of Changes*. New York: Ballantine Books. Translation of *Chou i* and related MSS from Ma-wang-tui.

————, ed. 1997b. *New Sources of Early Chinese History: An Introduction to the Reading of Inscriptions and Manuscripts.* Early China Special Monograph Series, 3. Berkeley: Society for the Study of Early China and the Institute of East Asian Studies, University of California.

Sinclair, R. K. 1988. *Democracy and Participation in Athens.* Cambridge University Press.

Singer, Charles. 1956. *Galen on Anatomical Procedures.* Oxford University Press.

Singer, Peter N. 1997. *Galen. Selected Works.* Oxford University Press.

Sivin, Nathan. 1966. Review of *The Book of Change*, translated by John Blofeld. *Harvard Journal of Asiatic Studies* 26: 290–98.

————. 1969. *Cosmos and Computation in Early Chinese Mathematical Astronomy.* Leiden: E. J. Brill.

————. 1978. On the Word "Taoism" as a Source of Perplexity, with Special Reference to the Relations of Science and Religion in Traditional China. *History of Religions* 17: 303–30. Reprinted in Sivin 1995e, chap. 6.

————. 1980. The Theoretical Background of Elixir Alchemy. In Needham 1954–, 5: pt. 4, 210–97.

————. 1987. *Traditional Medicine in Contemporary China. A Partial Translation of Revised Outline of Chinese Medicine* (1972) *with an Introductory Study on Change in Present-Day and Early Medicine.* Science, Medicine and Technology in East Asia, 2. Ann Arbor: Center for Chinese Studies, University of Michigan.

————. 1995a. The Myth of the Naturalists. In Sivin 1995e, chap. 4.

————. 1995b. Text and Experience in Classical Chinese Medicine. In Bates 1995, 177–204.

————. 1995c. On the Limits of Empirical Knowledge in Chinese and Western Science. In Sivin 1995e, chap. 5.

————. 1995d. *Science in Ancient China. Researches and Reflections.* Variorum Collected Studies Series. Aldershot, England: Variorum.

————. 1995e. *Medicine, Philosophy and Religion in Ancient China. Researches and Reflections.* Variorum Collected Studies Series. Aldershot, England: Variorum.

————. 1998. On the Dates of Yang Shang-shan and the *Huang-ti nei ching t'ai su. Chinese Science* 15: 29–36.

Sleeswyk, André Wegener, and Nathan Sivin. 1983. Dragons and Toads. The Chinese Seismoscope of A.D. 132. *Chinese Science* 6: 1–19.

Smith, Kidder, Jr., et al. 1990. *Sung Dynasty Uses of the I ching.* Princeton University Press.

Smith, Wesley D. 1973. Galen on Coans versus Cnidians. *Bulletin of the History of Medicine* 47: 569–85.

Staden, Heinrich von. 1989. *Herophilus: The Art of Medicine in Early Alexandria.* Cambridge University Press.

Sun Kuang-te 孫廣德. 1969. Hsien-Ch'in liang Han yin-yang wu-hsing shuo ti cheng-chih ssu-hsiang 先秦兩漢陰陽五行說的政治思想 (Political thought involving yin-yang and wu-hsing doctrines before and during the Han period). Chia-hsin Shui-ni Kung-ssu Wen-hua Chi-chin-hui yen-chiu lun-wen 嘉新水泥公司文化基金會研究論文, 147. Taipei: The Foundation.

Sun Xiaochun and Jacob Kistemaker. 1997. *The Chinese Sky during the Han. Constellating Stars and Society.* Leiden: E. J. Brill.

Temkin, Owsei. 1973. *Galenism. The Rise and Decline of a Medical Philosophy.* Ithaca: Cornell University Press.

Thivel, Antoine. 1981. *Cnide et Cos.* Paris: Les Belles Lettres.

Thomas, Rosalind. 1989. *Oral Tradition and Written Record in Classical Athens.* Cambridge University Press.

———. 1992. *Literacy and Orality in Ancient Greece.* Cambridge University Press.

Too, Yun Lee. 1995. *The Rhetoric of Identity in Isocrates: Text, Power, Pedagogy.* Cambridge University Press.

Toomer, Gerald J. 1984. *Ptolemy's Almagest.* London: Duckworth.

Tsien Tsuen-hsuin. 1985. *Chemistry and Chemical Technology. Paper and Printing.* In Needham 1954–, 5: pt. 1. Cambridge University Press.

Twitchett, Denis, and Michael Loewe, eds. 1986. *The Cambridge History of China.* Vol. 1, *The Ch'in and Han Empires, 221 B.C.–A.D. 220.* Cambridge University Press.

Vernant, Jean-Pierre. 1980. *Myth and Society in Ancient Greece.* Translated by Janet Lloyd. Hassocks, England: Harvester. Originally published as *Mythe et société en Grèce ancienne* (Paris: Maspero, 1974).

———. 1983. *Myth and Thought among the Greeks.* Translated by Janet Lloyd. London: Routledge. Originally published as *Mythe et pensée chez les grecs* (2d ed., Paris: Maspero, 1965).

Vernant, Jean-Pierre, and Jacques Gernet. 1964. L'évolution des idées en Chine et en Grèce du VIe au IIe siècle avant notre ère. *Bulletin de l'Association Guillaume Budé,* ser. 4, 3: 308–25. Translated by Janet Lloyd under the title "Social History and the Evolution of Ideas in China and Greece from the Sixth to the Second Centuries B.C." In Vernant 1980, 79–100.

Vidal-Naquet, Pierre. 1986. *The Black Hunter: Forms of Thought and Forms of Society in the Greek World*. Translated by A. Szegedy-Maszak. Baltimore: Johns Hopkins University Press. Originally published as *Le Chasseur noir* (Paris: La Découverte, 1981).

Vlastos, Gregory. 1975. *Plato's Universe*. Oxford: Clarendon.

Waerden, B. L. van der. 1960. Greek Astronomical Calendars and Their Relation to the Athenian Civil Calendar. *Journal of Hellenic Studies* 80: 168–80.

Walzer, Richard, and Michael Frede. 1985. *Galen, Three Treatises on the Nature of Science*. Indianapolis: Hackett.

Wang, Aihe. 2000. *Cosmology and Political Culture in Early China*. Cambridge Studies in Chinese History, 51. Cambridge University Press.

Wang Kuo-wei 王國維. 1927. *Kuan t'ang chi lin* 觀堂集林 (Collected works from the Hall of Contemplation). In *Hai-ning Wang Chung-ch'üeh-kung i-shu ch'u chi* 海寧王忠慤公遺書初集 (Surviving works of Wang Kuo-wei of Hai-ning, first collection). Peking: N.p. See the essay "Han Wei po-shih k'ao 漢魏博士考" (On Erudites in the Han and Wei periods).

Wang Li-ch'i 王利器. 1958. *Yen t'ieh lun chiao chu* 鹽鐵論校注 (Discourses on salt and iron, critically annotated). Shanghai: Ku-tien Wen-hsueh CPS.

Watson, Burton. 1958. *Ssu-ma Ch'ien. Grand Historian of China*. New York: Columbia University Press.

———. 1963. *Hsun-tzu. Basic Writings*. New York: Columbia University Press.

———. 1964. *Han Fei Tzu. Basic Writings*. Translations from the Oriental Classics. New York: Columbia University Press.

Wilhelm, Richard. 1950. *The I Ching or Book of Changes*. Translated by Cary F. Baynes. 2 vols. Bollingen Series, 19. New York: Pantheon.

Wolff, Michael. 1978. *Geschichte der Impetustheorie*. Frankfurt am Main: Suhrkamp.

Yamada Keiji 山田慶兒. 1979a. *Konton no umi e. Chūgokuteki shikō no kōzō* 混沌の海へ。中國的思考の構造 (Toward the sea of chaos. The structure of Chinese thought). Tokyo: Chikuma Shobō.

———. 1979b. The Formation of the *Huang-ti nei-ching*. *Acta Asiatica* 36: 67–89.

———. 1980. *Kyukū hachifū setsu to Shoshiha no tachiba* 九宮八風説と少師派の立場 (The nine palaces–eight winds doctrine and the standpoint of the Shao shih lineage). *Tōhō gakuhō* (Kyoto) 52: 199–242.

———. 1991. Anatometrics in Ancient China. *Chinese Science* 10: 39–52.

Yang, Hsien-i, and Gladys Yang, trans. 1979. *Selections from Records of the Historian*. Peking: Foreign Languages Press. Translations from *Shih chi*.

Yates, Robin D. S. 1975. Towards a Reconstruction of the Tactical Chapters of the *Mo tzu (chüan* 14). M.A. thesis, University of California, Berkeley.

————. 1997. *Five Lost Classics: Tao, Huang-Lao, and Yin-Yang in Han China*. New York: Ballantine Books.

Yü Shu-lin 余書麟. 1966. Liang Han ssu hsueh yen-chiu 兩漢私學研究 (A study of private academies in the Han). *Shih-ta hsueh-pao* 師大學報 11. 1: 109–47.

Yü Ying-shih 余英時. 1980. Ku-tai chih-shih chieh-ts'eng ti hsing-ch'i yü fa-chan 古代知識階層的興起與發展 (The rise and development of the intellectual stratum in ancient times). In *Chung-kuo chih-shih chieh-ts'eng shih lun. Ku-tai p'ien*. 中國知識階層史論。古代篇 (Historical discussions of the intellectual stratum in China. Ancient China), 1–108. Taipei: Lien Ching.

Zufferey, Nicolas. 1998. Erudits et lettrés au début de la dynastie Han. *Asiatische Studien* 52. 3: 915–65.

Zürcher, Erik. 1980. Buddhist Influence on Early Taoism. A Survey of Scriptural Evidence. *T'oung Pao* 66: 84–147.

Index and Glossary

This index includes important persons, with brief information about their careers and dates. Where birth and death dates are not both known, "b." designates birth and "d." death. In listings of rulers, "r." stands for "reigned." In approximate dates, "fl." stands for "flourished." The index also lists texts, with titles of those that form part of an extant book in quotation marks. Books with more than one title are listed under the earliest (e.g., Chou pi 周髀 rather than Chou pi suan ching 周髀算經). We note the main meanings of the most important Greek and Chinese concepts. For a guide to using the Pinyin system to look up items in this index, which employs the Wade-Giles system, see the most significant differences between the two listed in the entry "romanization, Pinyin, with Wade-Giles equivalents."